Genomic Medicine

For Kelly —

With thanks for all that you have taught me about genetics.

— Alan

Genomic Medicine

Articles from the

New England Journal of Medicine

Edited by

ALAN E. GUTTMACHER, M.D.

FRANCIS S. COLLINS, M.D., Ph.D.

JEFFREY M. DRAZEN, M.D.

With a Foreword by

ELIAS ZERHOUNI, M.D.

THE JOHNS HOPKINS UNIVERSITY PRESS
Baltimore & London

THE NEW ENGLAND JOURNAL OF MEDICINE
Boston

Disclaimer: Reference to the National Institutes of Health (NIH) should not be viewed as an endorsement or implied endorsement of the *New England Journal of Medicine* or the publisher, their goods, or services. The opinions of the authors should not be viewed as those of the NIH.

Printed in the United States of America on acid-free paper
2 4 6 8 9 7 5 3 1

The Johns Hopkins University Press
2715 North Charles Street
Baltimore, Maryland 21218-4363
www.press.jhu.edu

Library of Congress Cataloging-in-Publication Data

Genomic medicine : articles from the New England Journal of Medicine / edited by Alan E. Guttmacher, Francis S. Collins, and Jeffrey M. Drazen; with a foreword by Elias Zerhouni.
p. ; cm.
Includes bibliographical references and index.
ISBN 0-8018-7979-5 (pbk. : alk. paper)
1. Medical genetics. 2. Gene therapy. 3. Genetic screening. 4. Genomics.
[DNLM: 1. Genetic Screening—Collected Works. 2. Gene Therapy—Collected Works. 3. Genomics—Collected Works. 4. Molecular Diagnostic Techniques—Collected Works. 5. Pharmacogenetics—Collected Works. QZ 50 G3355 2004] I. Guttmacher, Alan E. II. Collins, Francis S. III. Drazen, Jeffrey M., 1946– IV. New England Journal of Medicine.
RB155.G463 2004
616'.042—dc22

2003021855

A catalog record for this book is available from the British Library.

Contents

Foreword

Fifteen years ago, many questioned the wisdom of sequencing the human genome. There were concerns not only about the technical feasibility of such an effort but also that its scientific worth would prove too modest to warrant the investment of human and financial capital that the project would require. This collection of essays stands as eloquent testimony to those who overcame such doubts to create, nurture, and finish the sequencing of the human genome in April 2003, a date that will forever stand out in human history. The wide and public availability of the human genome sequence and the other tools spawned by the Human Genome Project have helped to create an unparalleled era of biomedical discovery. The project has opened up new avenues for the understanding of the biology of human health and disease. Many of these avenues remain to be fully explored, but this volume already demonstrates that information about genetics and genomics is increasingly permeating our clinical approach to patients.

A central challenge for medicine in this century will be to evolve from the secular model of intervention *after* loss of normal function to more predictive models of intervention *before* the onset of disease. Key to this shift is our unfolding knowledge of the genetic factors in health and disease gained from the Human Genome Project and their translation into predictable clinical phenotypes in the context of defined environmental drivers. As several of the chapters in this book document, we are already gaining profound new insights into the basic biology of health and disease (even diseases with relatively little genetic input) that may lead directly to new diagnostic, preventive, and therapeutic strategies. It would not surprise me if new and revolutionary taxonomic models of disease were to evolve in the not too distant future through the combination of astute clinical and genomic observations and analyses. One can only imagine the profound institutional consequences this would have.

It is important to remember, however, that genomics is still a very young field; the very word *genomics* is less than twenty years old. It is obvious, but worth emphasizing, that we are extremely early in what promises to be the long history of the application of genomics to clinical care. For instance, genomics today contributes very modestly to the routine care of a pregnant patient. Yet, by the time the child born of that pregnancy reaches adulthood, it is likely that knowledge of his or her genetic makeup will not just help guide the treatment of most health problems but help prevent, or at least postpone or lessen, the very manifestations of many of them.

The full blossoming of the era of "personalized medicine" still lies a few years ahead. It will take substantial and varied translational and clinical research to convert this concept from a convenient catch phrase to an effective reality. Such research will occur only if health care providers from many disciplines become knowledgeable about genomics and play a leading part in designing and carrying it out. Similarly, the widespread incorporation of the results of such research into clinical practice will require a clinical workforce well informed about genomics and able to utilize it easily and effectively. It is only the clinical researcher steeped in genomics who can develop the applications of genomics to health care, and it is only the clinician comfortable with genomics who can implement them. Just as important will be the necessary and timely incorporation of this knowledge into the curricula of all of our professional schools.

This book also demonstrates that one of the unusual features of the Human Genome Project—its devotion of some of its funds to research into the ethical, legal, and social implications of science—has already paid important dividends. There now exists a rich body of scholarship, from a variety of perspectives, about the societal implications of genomics. This is important because the increased clinical use of genomics will bring issues at the interface of genomics and society to the fore. Among these are the potential for genetics-based discrimination (especially in insurance and employment but also in areas such as education), questions about equitable access to genomics-based care, the application of genomic medicine in the developing world, intellectual property issues, and new perspectives on race and ethnicity. Issues as significant as these will also demand thoughtful research that involves a wide variety of expertise.

The Human Genome Project has begun to change clinical care; it has also changed the conduct of research. The project demonstrated that a model of large but coordinated inquiry more typical of the physical sciences can be successful in the biomedical sciences as well. It also showed that research co-supported and co-led by different branches of government (in this case, the U.S. Department of Energy and the National Institutes of Health in the Department of Health and Human Services) can be successful, just as it showed that science can be benefited by truly international collaboration. The project demonstrated how biomedical research can spur technology development and in so doing become a driver of economic development. Its emphasis on immediate, free, and open release of data has helped this ideal permeate other areas of pre-competitive research. Furthermore, the Human Genome Project has reminded us that exciting and productive research can interest large segments of society in science and inspire and attract talented young people to scientific and medical careers.

I am proud that the National Institutes of Health has had a key role in supporting the creation and early development of the still young field of genomics, and I pledge that we will continue to support its maturation vigorously. This volume and the series in the *New England Journal of Medicine* that inspired it are important and authoritative surveys of the genomic medicine of today. However, I anticipate eagerly the day, still in my professional lifetime, that the *New England Journal of Medicine* publishes a new series on genomic medicine that dwarfs this volume in its scope and depth, thus making the book you are holding an antiquated, if historic, attempt to grapple with the complex application of genomics to medicine. I commend the authors of this book and look forward to the creative energy of the many researchers and clinicians who will develop the knowledge in that yet-to-be-published volume.

—ELIAS ZERHOUNI, M.D.
Director, National Institutes of Health

Preface

When I became Editor-in-Chief of the *New England Journal of Medicine* in July of 2000, one of my key objectives was to raise the awareness of physicians about the recent advances in genetics and genomics through a series of articles that highlighted how this new knowledge was changing the face of medicine. I had hoped to recruit Francis Collins, director of the National Human Genome Research Institute, as the series editor. To my surprise, before I had a chance to write, Francis Collins and Alan Guttmacher wrote to me with the same idea. We met in Bethesda a few weeks later, and we soon had an outline for the Genomic Medicine series that ran in the *Journal* from November 2002 through September 2003. Our premise was that there had been too much "we will be able to do this and that" and not enough about what genetics and genomics had already brought to our understanding of the biology of disease and the practice of medicine. The series encapsulated in this volume is a snapshot of how genetics and genomics have advanced our understanding of disease biology, helped in making specific diagnoses, and led to novel and specific treatments. I urge savvy clinicians to read it soon because the field is unfolding quickly, and we will see an exciting evolutionary leap in medicine over the next decade.

—JEFFREY M. DRAZEN, M.D.
Parker B. Francis Professor of Medicine
at Harvard Medical School
Chief of Pulmonary and Critical Care Medicine
at Brigham and Women's Hospital

Contributors

WYLIE BURKE, M.D., Ph.D.
Department of Medical History and Ethics
University of Washington
Seattle, Washington

ELLEN WRIGHT CLAYTON, M.D., J.D.
Center for Genetics and Health Policy
Vanderbilt University
Nashville, Tennessee

FRANCIS S. COLLINS, M.D., Ph.D.
National Human Genome Research Institute
National Institutes of Health
Bethesda, Maryland

ALBERT DE LA CHAPELLE, M.D., Ph.D.
Human Cancer Genetics Program
Comprehensive Cancer Center
Ohio State University
Columbus, Ohio

CHRISTOPHER E. ELLIS, Ph.D.
Genetic Diseases Research Branch
National Human Genome Research Institute
National Institutes of Health
Bethesda, Maryland

WILLIAM E. EVANS, Pharm.D.
St. Jude Children's Research Hospital
and University of Tennessee Colleges of Pharmacy and Medicine
Memphis, Tennessee

DAVID B. GOLDSTEIN, Ph.D.
Department of Biology
University College London
London, England

ALAN E. GUTTMACHER, M.D.
National Human Genome Research Institute
National Institutes of Health
Bethesda, Maryland

MUIN J. KHOURY, M.D., Ph.D.
Office of Genomics and Disease Prevention
Centers for Disease Control and Prevention
Atlanta, Georgia

HENRY T. LYNCH, M.D.
Department of Preventive Medicine and Public Health
Creighton University School of Medicine
Omaha, Nebraska

EDWARD R. B. McCABE, M.D., Ph.D.
Department of Pediatrics
David Geffen School of Medicine at UCLA
Los Angeles, California

LINDA L. McCABE, Ph.D.
Department of Human Genetics
David Geffen School of Medicine at UCLA
and UCLA Center for Society, the Individual and Genetics
Los Angeles, California

HOWARD L. McLEOD, Pharm.D.
Washington University Medical School
Saint Louis, Missouri

ELIZABETH G. NABEL, M.D.
National Heart, Lung, and Blood Institute
National Institutes of Health
Bethesda, Maryland

ROBERT L. NUSSBAUM, M.D.
Genetic Diseases Research Branch
National Human Genome Research Institute
National Institutes of Health
Bethesda, Maryland

LOUIS M. STAUDT, M.D., Ph.D.
Metabolism Branch
Center for Cancer Research
National Cancer Institute
Bethesda, Maryland

HAROLD VARMUS, M.D.
Memorial Sloan-Kettering Cancer Center
New York, New York

BARBARA L. WEBER, M.D.
University of Pennsylvania
Philadelphia, Pennsylvania

RICHARD WEINSHILBOUM, M.D.
Department of Molecular Pharmacology, Mayo Medical School
Department of Experimental Therapeutics, Mayo Clinic
Department of Medicine, Mayo Foundation
Rochester, Minnesota

RICHARD WOOSTER, Ph.D.
Wellcome Trust Sanger Institute
Hinxton, Cambridge
United Kingdom

Glossary

The following terms are used in this book. (For a "talking glossary" of many genetics terms, see http://www.genome.gov/glossary.cfm.)

Allele—An alternative form of a gene.

Alternative splicing—A regulatory mechanism by which variations in the incorporation of a gene's exons, or coding regions, into messenger RNA lead to the production of more than one related protein, or isoform.

Autosomes—All of the chromosomes except for the sex chromosomes and the mitochondrial chromosome.

Centromere—The constricted region near the center of a chromosome that has a critical role in cell division.

Codon—A three-base sequence of DNA or RNA that specifies a single amino acid.

Conservative mutation—A change in a DNA or RNA sequence that leads to the replacement of one amino acid with a biochemically similar one.

Epigenetic—A term describing nonmutational phenomena, such as methylation and histone modification, that modify the expression of a gene.

Exon—A region of a gene that codes for a protein.

Frame-shift mutation—The addition or deletion of a number of DNA bases that is not a multiple of three, thus causing a shift in the reading frame of the gene. This shift leads to a change in the reading frame of all parts of the gene that are downstream from the mutation, often leading to a premature stop codon and, ultimately, to a truncated protein.

Gain-of-function mutation—A mutation that produces a protein that takes on a new or enhanced function.

Genomics—The study of the functions and interactions of all the genes in the genome, including their interactions with environmental factors.

Genotype—A person's genetic makeup, as reflected by his or her DNA sequence.

Haplotype—A group of nearby alleles that are inherited together.

Heterozygous—Having two different alleles at a specific autosomal (or X chromosome in a female) gene locus.

Homozygous—Having two identical alleles at a specific autosomal (or X chromosome in a female) gene locus.

Intron—A region of a gene that does not code for a protein.

Linkage disequilibrium—The nonrandom association in a population of alleles at nearby loci.

Loss-of-function mutation—A mutation that decreases the production or function of a protein (or does both).

Missense mutation—Substitution of a single DNA base that results in a codon that specifies an alternative amino acid.

Monogenic—Caused by a mutation in a single gene.

Motif—A DNA-sequence pattern within a gene that, because of its similarity to sequences in other known genes, suggests a possible function of the gene, its protein product, or both.

Multifactorial—Caused by the interaction of multiple genetic and environmental factors.

Nonconservative mutation—A change in the DNA or RNA sequence that leads to the replacement of one amino acid with a very dissimilar one.

Nonsense mutation—Substitution of a single DNA base that results in a stop codon, thus leading to the truncation of a protein.

Penetrance—The likelihood that a person carrying a particular mutant gene will have an altered phenotype.

Phenotype—The clinical presentation or expression of a specific gene or genes, environmental factors, or both.

Point mutation—The substitution of a single DNA base in the normal DNA sequence.

Regulatory mutation—A mutation in a region of the genome that does not encode a protein but affects the expression of a gene.

Repeat sequence—A stretch of DNA bases that occurs in the genome in multiple identical or closely related copies.

Silent mutation—Substitution of a single DNA base that produces no change in the amino acid sequence of the encoded protein.

Single-nucleotide polymorphism (SNP)—A common variant in the genome sequence; the human genome contains about 10 million SNPs.

Stop codon—A codon that leads to the termination of a protein rather than to the addition of an amino acid. The three stop codons are TGA, TAA, and TAG.

Genomic Medicine

1

Getting Ready for Gene-Based Medicine

HAROLD VARMUS, M.D.

Despite much journalistic hyperbole, the publication in 2001 of nearly complete sequences of the human genome[1,2] did not mean that the practice of medicine would be abruptly and radically transformed. Medicine has not been a gene-free art form in living memory—a knowledge of genetics has had an increasingly important role in medicine for over a century. Nor will it soon, if ever, become an impersonal, information-based science enabling every person to know from birth the ailments he or she will have and the perfect way to treat them.

Still, changes in medical practice are already occurring at an accelerating pace under the influence of the elucidation of genomes. Medical genetics, once a tool for diagnosing a handful of relatively rare diseases inherited in a simple mendelian fashion, has expanded into new territories: the prediction of a healthy person's risks of even common diseases such as cancer and cardiovascular disease; the analysis of patterns of gene expression as an adjunct to conventional diagnostic methods, such as histopathology; and the evaluation of multigenic diseases and responses to environmental agents and drugs. Knowledge about the genomes of microbes is expanding the opportunities for diagnosing, preventing, and treating infectious diseases, and it is likely that such knowledge will soon contribute to defending our nation against bioterrorism. But the full potential of a DNA-based transformation of medicine will be realized only gradually, over the course of decades, as we try to understand the content of genomes and, most important, the physiological consequences of variations in their sequence.

The pace of this transformation will be limited not only by the pace of discovery, but also by the need to educate practicing physicians, their coworkers, and their patients about the uses and shortcomings of genetic information. Unfortunately, most medical schools did not anticipate the changes that molecular genetics would bring to modern medicine. As a result, the ranks of medical geneticists are sparse, and many physicians struggle with

Originally published November 7, 2002

the new biology. Furthermore, the nation's battalion of genetic counselors has never grown to the size that would be needed in order to compensate for these deficiencies. As a result, doctors, nurses, and the public will have to do some work on their own to learn about the genes and genomes that will progressively change medical practice.

These changes in medicine are particularly gratifying to those of us who have devoted most of our professional lives to deciphering gene-based mechanisms of disease in the hope of improving health. But one message that emerges from the need for this book is the imperative to do a better job of anticipating and planning for all the consequences of such profound changes. If we are to provide the public with the full benefits of medical science, we must now begin to ask some troubling questions: How can we better train the next generations of physicians to practice genetic medicine? How can increasingly complex genetic knowledge be made readily accessible to all practitioners when they need it? How much will the expanded use of gene-based methods further escalate the cost of health care, and who will pay for it? How can we ensure that these products of our science, largely financed by federal dollars, will reach all the citizens of our country? If this book refocuses attention on these vexing social issues, it will serve a function just as important as the mission of teaching doctors, nurses, and patients the state of the science.

References

1. International Human Genome Sequencing Consortium. Initial sequencing and analysis of the human genome. Nature 2001; 409:860–921.

2. Venter JC, Adams MD, Myers EW, et al. The sequence of the human genome. Science 2001; 291: 1304–51.

2

Genomic Medicine—A Primer

ALAN E. GUTTMACHER, M.D., and FRANCIS S. COLLINS, M.D., Ph.D.

Humans have known for millennia that heredity affects health.[1] However, Mendel's seminal contribution to the elucidation of the mechanisms by which heredity affects phenotype occurred less than 150 years ago, and Garrod began applying this knowledge to human health only at the start of the past century. For most of the 20th century, many medical practitioners viewed genetics as an esoteric academic specialty; that view is now dangerously outdated.

THE ADVENT OF GENOMIC MEDICINE

The recent completion of the draft sequence of the human genome[2,3] and related developments have increased interest in genetics, but confusion remains among health professionals and the public about the role of genetic information in medical practice. Inaccurate beliefs about genetics persist, including the view that in the past it had no effect on the practice of medicine and that its influence today is pervasive. In fact, for decades knowledge of genetics has had a large role in the health care of a few patients and a small role in the health care of many. We have recently entered a transition period in which specific genetic knowledge is becoming critical to the delivery of effective health care for everyone.

If genetics has been misunderstood, genomics is even more mysterious—what, exactly, is the difference? Genetics is the study of single genes and their effects. "Genomics,"[4] a term coined only about 15 years ago, is the study not just of single genes, but of the functions and interactions of all the genes in the genome. Genomics has a broader and more ambitious reach than does genetics. The science of genomics rests on direct experimental access to the entire genome and applies to common conditions, such as breast cancer[5] and colorectal cancer,[6] human immunodeficiency virus (HIV) infection,[7] tuberculosis,[8]

Originally published November 7, 2002

Parkinson's disease,[9] and Alzheimer's disease.[10] These common disorders are also all due to the interactions of multiple genes and environmental factors. They are thus known as multifactorial disorders. Genetic variations in these disorders may have a protective or a pathologic role in the expression of diseases.

The role of genomics in health care is in part highlighted by the decreasing effect of certain environmental factors, such as infectious agents, on the burden of disease. Genomics also contributes to the understanding of such important infectious diseases as the acquired immunodeficiency syndrome (AIDS)[7] and tuberculosis.[8]

The following two case vignettes illustrate how knowledge of genomics may lead to better management of common medical conditions.

Thirty-four-year-old Kathleen becomes pregnant and sees a new physician for her first prenatal visit. Her medical history is remarkable for an episode of deep venous thrombosis five years earlier while she was taking oral contraceptives; her mother had had deep venous thrombosis when pregnant with Kathleen. Her physician suspects that Kathleen has a hereditary thrombophilia and obtains blood tests to screen for a genetic predisposition to thrombosis. Kathleen proves to be among the approximately 4 percent of Americans who are heterozygous for a mutation in factor V known as factor V Leiden that increases the risk of thrombotic events. On the basis of this knowledge and her history of possibly estrogen-related thromboembolism, she is treated with prophylactic subcutaneous heparin for the balance of her pregnancy. She remains asymptomatic and delivers a healthy, term infant.

Four-year-old John has acute lymphoblastic leukemia and tolerates induction and consolidation chemotherapy well, with minimal side effects. As a key part of his maintenance-treatment protocol, he begins to receive oral mercaptopurine daily, but because a genetic test shows that John is homozygous for a mutation in the gene that encodes thiopurine S-methyltransferase, an enzyme that inactivates mercaptopurine, he receives a greatly reduced dose. Only a few years ago, about 1 in 300 patients had serious, sometimes lethal, hematopoietic adverse effects during mercaptopurine therapy. Although John is in this at-risk minority, a simple genetic test, which is now routine for patients beginning mercaptopurine therapy, alerts his physicians to this genetic predisposition. They reduce his dose of mercaptopurine and carefully monitor his blood levels, ensuring that the drug levels remain therapeutic, rather than toxic. John subsequently has an uneventful several-year maintenance period and achieves complete remission.

THE HUMAN GENOME

These are two examples of genomic medicine, the application of our rapidly expanding knowledge of the human genome (Fig. 2.1) to medical practice. Much is known, but much remains mysterious. We know that less than 2 percent of the human genome codes for proteins, while over 50 percent represents repeat sequences of several types, whose function is less well understood. These stretches of repetitive sequences, sometimes wrongly dismissed as "junk DNA," constitute an informative historical record of evolutionary biology, provide a rich source of information for population genetics and medical genetics, and by introducing changes into coding regions, are active agents for change within the genome.[2]

A draft sequence of the human genome was announced in 2000, and by September 2002 over 90 percent was in final form—that is, there were no gaps and the data were greater than 99.99 percent accurate. The entire 3.1 gigabases of the DNA sequence were essentially completed (except for the centromeres and rare unclonable segments) in April 2003. Even with this knowledge in hand, however, we still do not know precisely how many genes the genome contains. Current data indicate that the human genome includes approximately 30,000 to 35,000 genes[2]—a number that is substantially smaller than was previously thought. Only about half these genes have recognizable motifs, or DNA-sequence patterns, that suggest possible functions. Mutations known to cause disease have been identified in approximately 1000 genes. However, it is likely that nearly all human genes are capable of causing disease if they are altered substantially. Whereas it was once dogma that one gene makes one protein, it now appears that, through the mechanism of alternative splicing[12] (Fig. 2.2), more than 100,000 proteins can be derived from these 30,000 to 35,000 genes.[2]

In addition to alternative splicing, a number of "epigenetic" phenomena, such as methylation and histone modification, can alter the effect of a gene. Furthermore, a complex array of molecular signals allows specific genes to be "turned on" (expressed) or "turned off" in specific tissues and at specific times.

Genes are distributed unevenly across the human genome (Fig. 2.1). Certain chromosomes, particularly 17, 19, and 22, are relatively gene dense as compared with others, such as 4, 8, 13, 18, and Y.[3] Moreover, gene density varies within each chromosome, being highest in areas rich in the bases cytosine and guanine, rather than adenine and thymine.[2,3] Chromosomes 13, 18, and 21, the three autosomes with the fewest genes, are also the three for which the occurrence of trisomy (i.e., three copies of a chromosome) is compatible with viability.

Not all genes reside on nuclear chromosomes; several dozen involved with energy metabolism are on the mitochondrial chromosome.[13] Since ova are rich in mitochondria and sperm are not, mitochondrial DNA is usually inherited from the mother. Therefore, mitochondrial genes—and diseases due to DNA-sequence variants in them—are transmitted in a matrilineal pattern that is distinctly different from the pattern of inheritance of nuclear genes.

MONOGENIC CONDITIONS

Over the course of the 20th century, a combination of theoretical insights, basic-science research, and clinical observation elucidated the inheritance of single-gene, or monogenic, disorders (also known as mendelian disorders, since they are transmitted in a manner consonant with Mendel's laws of inheritance). Modes of inheritance have been established for thousands of conditions caused by mutations in single genes; these have been catalogued in a textbook[11] and, more recently, in an online compendium[14] known as Mendelian

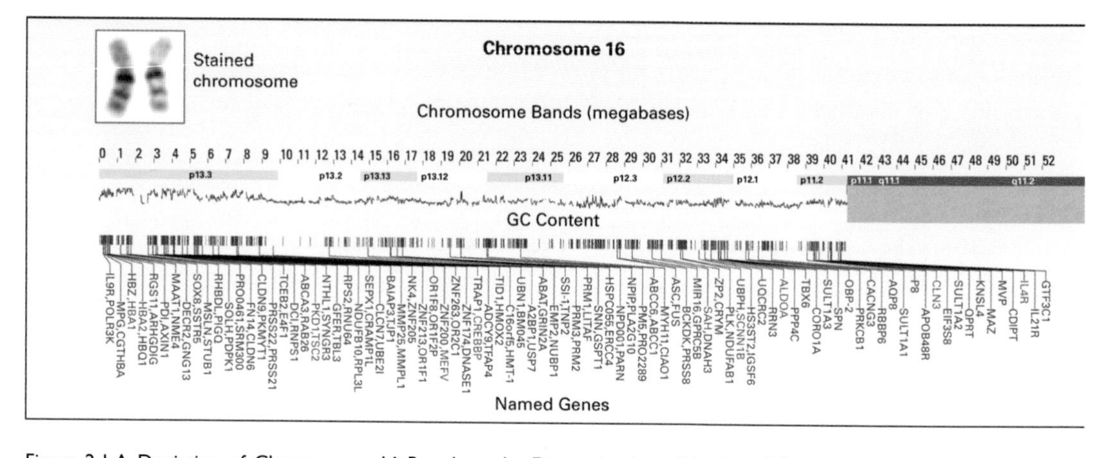

Figure 2.1 A Depiction of Chromosome 16 Based on the Determination of Its Actual Sequence by the Human Genome Project. The inset shows a Giemsa-stained chromosome. The figure shows a scale in megabases (1 megabase equals 1 million base pairs); the approximate positions of Giemsa-stained chromosome bands at a resolution of 800 bands; the proportion of bases in a 20,000-base window that are either guanine or cytosine (the GC content); the location of predicted genes; and the locations of named genes that have been located in the draft sequence (known disease genes in the Online Mendelian Inheritance

Inheritance in Man (OMIM). For nearly 100 years, autosomal dominant, autosomal recessive, and X-linked modes of inheritance have been understood and known to cause human disease. In the past few decades, other mechanisms of monogenic inheritance have been described. These include mitochondrial inheritance,[13] imprinting (a mechanism by which the effects of certain genes depend on whether they are inherited through the mother or through the father),[15] uniparental disomy (the occasional situation in which both members of one pair of a person's 23 pairs of chromosomes derive from one parent),[16] and expanding trinucleotide repeats (a phenomenon in which a sequence of three base pairs that is normally repeated a number of times in a row in the genome becomes repeated by more than the normal number of times, sometimes causing disease).[17]

Most single-gene conditions are uncommon. Even the commonest, such as hereditary hemochromatosis (approximate incidence, 1 in 300 persons), cystic fibrosis (approximate incidence, 1 in 3000), alpha$_1$-antitrypsin deficiency (approximate incidence, 1 in 1700), or neurofibromatosis (approximate incidence, 1 in 3000), affect no more than 1 in several hundred people in the United States. However, the total effect of monogenic conditions is substantial, from both the individual patient's and public health perspectives, and increased understanding of genetics has already begun to improve the health of some patients with such conditions. The delineation of the mechanisms by which genetic factors cause monogenic disorders has provided important information about basic pathophysiological processes that underlie related disorders that occur with far greater frequency than do these genetic disorders. For instance, insights regarding familial hypercholesterolemia, a genetic disorder that affects only 1 of every 500 people in the United States, were instrumental to understanding the pathophysiology of atherosclerosis, which affects a large fraction of the population, and the development of the statin drugs, which are among the most frequently prescribed medications.[18]

in Man [OMIM] data base[11] are in red, and other genes in blue). Only about 30 percent of human genes have been named thus far. There is considerable variation in gene density across the chromosome, particularly in the dark band 16q21 on the right, which has an extremely low gene density and GC content. The rectangular dark-gray block present on both pages represents the centromere and adjacent repetitive sequence on 16q. Modified from the work of the International Human Genome Sequencing Consortium.[2]

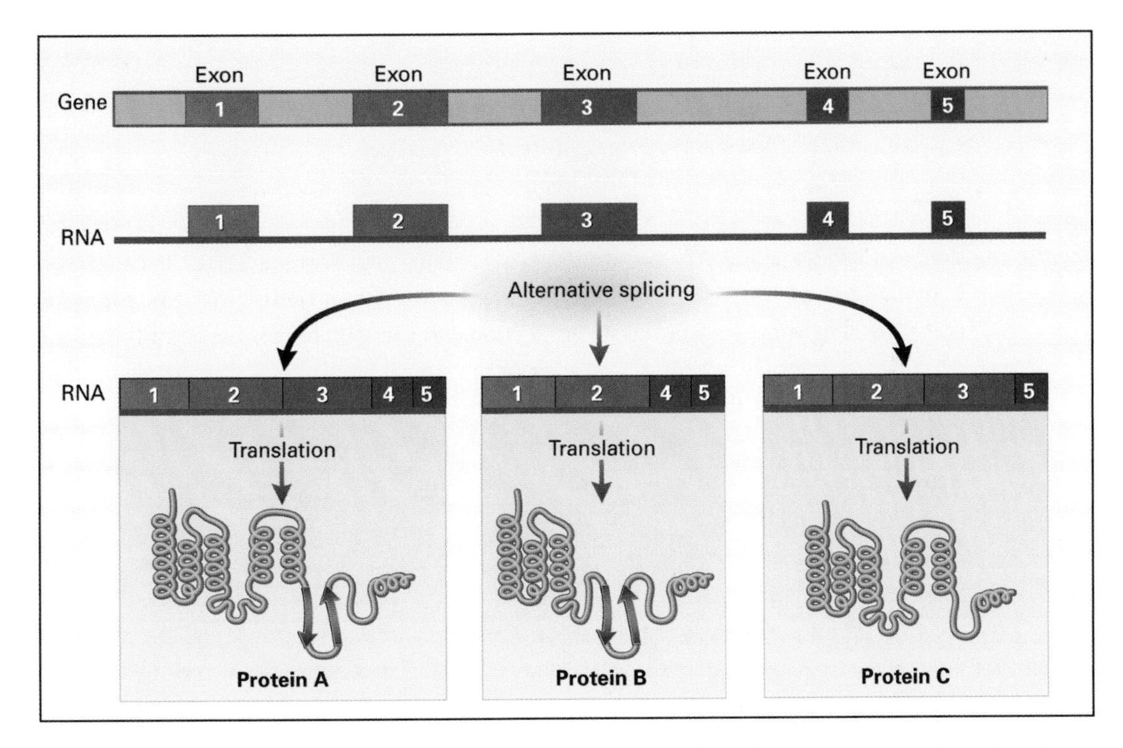

Figure 2.2 Alternative Splicing. A single gene can produce multiple related proteins, or isoforms, by means of alternative splicing.

TYPES OF MUTATION

There are a number of ways to categorize mutations. One is according to the causative mechanism, whereas another is according to their functional effect. When classified according to the mechanism, point mutations—that is, a change in a single DNA base in

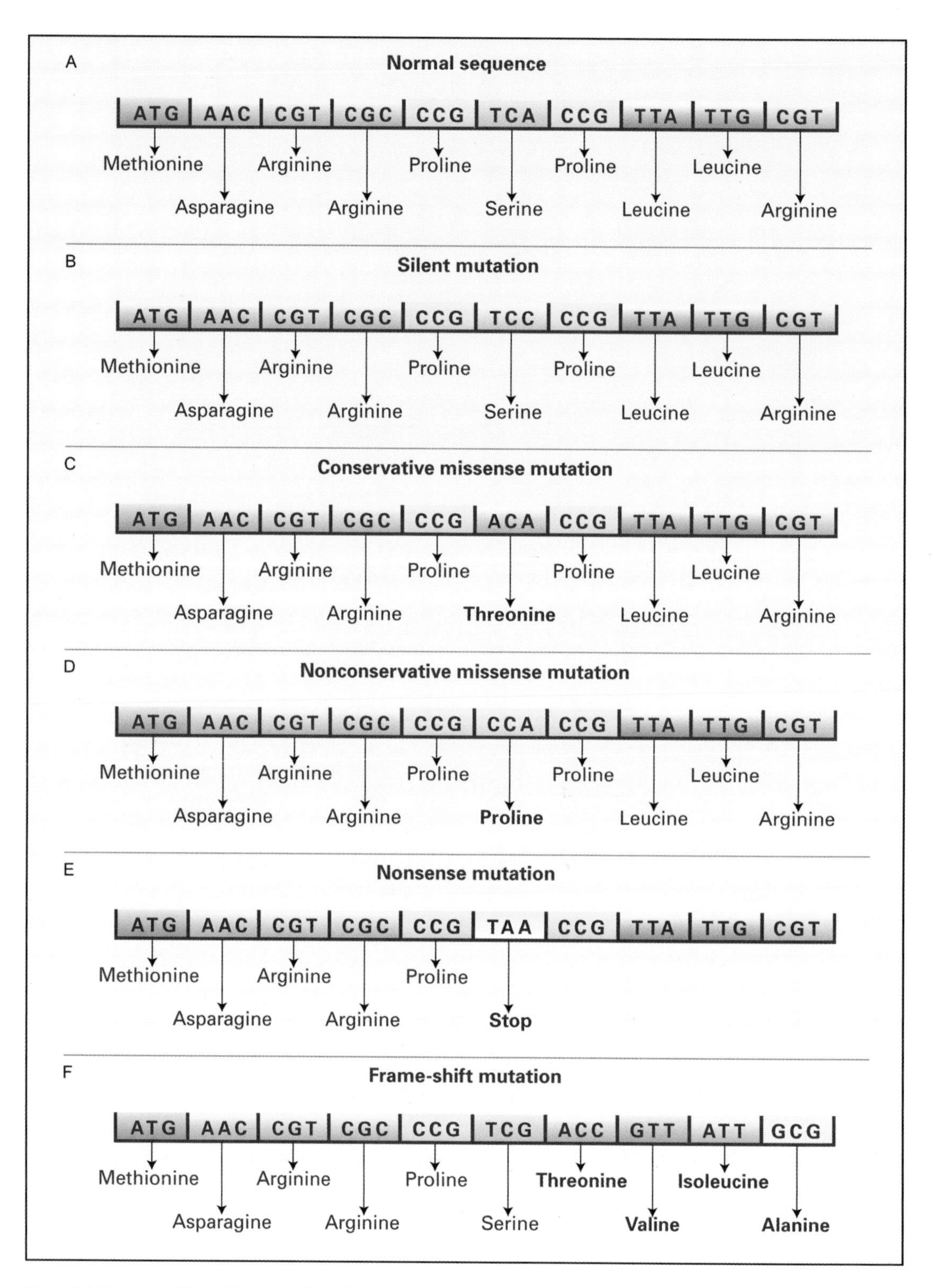

Figure 2.3 Examples of Point Mutations. Panel A shows the normal sequence of DNA from one exon and the protein product it encodes. Panel B shows a silent mutation, Panel C a conservative missense mutation (serine and threonine have very similar structures), Panel D a nonconservative missense mutation (serine and proline have very different structures), Panel E a nonsense mutation, and Panel F a frame-shift mutation. In Panel F, the insertion of a single G throws off the reading frame, so that all amino acids downstream are changed radically.

the sequence—are the most common. There are many types of point mutations. One type is a missense mutation (Fig. 2.3), a substitution that leads to an alternative amino acid, because of the way in which it changes the three-base sequence, or codon, that codes for an amino acid. Nonsense mutations (Fig. 2.3) are a more dramatically deleterious type of point mutation that change the codon to a "stop" codon, a codon that causes the termination of the protein instead of producing an amino acid. Another type of mutation is the frame-shift mutation (Fig. 2.3), which changes the reading frame of the gene downstream from it, often leading to a premature stop codon.

In terms of functional effect, rather than mechanism, many variants in the human-genome sequence have no phenotypic effect. Among these are silent mutations (Fig. 2.3), which replace one base with another, so that the resultant codon still codes for the same amino acid. Also, mutations may not change the phenotype if the altered codon substitutes one amino acid for another that produces little change in the function of the protein or proteins that the gene encodes. These are referred to as "conservative mutations" (Fig. 2.3). Nonconservative mutations (Fig. 2.3) replace an amino acid with a very different one and are more likely to affect the phenotype.

Although mutations can cause disease by a variety of means, the most common is loss of function. Loss-of-function mutations alter the phenotype by decreasing the quantity or the functional activity of a protein. For instance, mutations in the glucose-6-phosphate dehydrogenase *(G6PD)* gene on the X chromosome decrease the functional activity of this enzyme, leading to acute hemolytic anemia if a male (who would, of course, have only one copy of the X chromosome) with the mutation is exposed to certain drugs, including sulfonamides, primaquine, and nitrofurantoins. Since genes do not exist just to handle pharmacologic agents, variants that cause a more severe deficiency of glucose-6-phosphate dehydrogenase also lead to hemolytic anemia when affected males ingest fava beans (favism), since this enzyme is also important in the degradation of a component of the beans.[19]

Some mutations cause disease through a gain of function, whereby the protein takes on some new, toxic function. Expanding exonic CAG trinucleotide repeats that cause disorders including Huntington's disease and spinocerebellar ataxia appear to lead to neuropathologic abnormalities by producing proteins that function abnormally because of expanded polyglutamine tracts (CAG codes for the amino acid glutamine).[20] Such gain-of-function mutations are often dominantly inherited, since a single copy of the mutant gene can alter function.

One might assume that mutations in the approximately 98.5 percent of the genome that does not code for proteins do not affect the phenotype. Indeed, most do not. But others are regulatory mutations that may ultimately prove as important in the etiologic process of common diseases as the coding region variants. Regulatory mutations act by altering the expression of a gene. For instance, a regulatory mutation might lead to the loss of expression of a gene, to unexpected expression in a tissue in which it is usually silent, or to a change in the time at which it is expressed. Examples of regulatory mutations associated with disease are those in the flanking region of the *FMR1* gene (causing fragile X syndrome),[21] the insulin gene flanking region (increasing the risk of type 1 diabetes mellitus),[22] a regulatory site of the type I collagen gene (increasing the risk of osteoporosis),[23] and an intronic regulatory site of the calpain-10 gene (increasing the risk of type 2 diabetes mellitus).[24]

Mutations can also decrease the risk of a disease. One example of this is a 32-bp deletion (a frame shift) in a chemokine receptor gene, *CCR5*. Persons who are homozygous for this deletion prove almost completely resistant to infection with HIV type 1, and those who are heterozygous for the deletion have slower progression from infection to AIDS. These effects arise because *CCR5* is an important part of the mechanism by which HIV enters the cell.[25]

GENES IN COMMON DISEASE

The study of genomics will most likely make its greatest contribution to health by revealing mechanisms of common, complex diseases, such as hypertension, diabetes, and asthma. So far, most genes involved in common diseases have been identified by virtue of their high penetrance—that is, the mutations lead to disease in a fairly large proportion of people who have them. Examples include mutations in *BRCA1* and *BRCA2* (increasing the risk of breast and ovarian cancer),[26] *HNPCC* (increasing the risk of hereditary nonpolyposis colorectal cancer),[6] *MODY 1*, *MODY* 2, and *MODY* 3 (increasing the risk of diabetes),[27] and the gene for α-synuclein (causing Parkinson's disease).[9] One can think of these as near-mendelian subgroups of disease within a larger group of affected persons. If a person has such a mutation, the likelihood of disease is great. However, each of these highly penetrant mutations associated with common disease has a prevalence in the general population of only one in several hundred to several thousand people.

From a public health perspective, genes with mutations that are less highly penetrant but much more prevalent have a greater effect on the population than genes that are highly penetrant but uncommon. Such mutations have been reported in genes such as *APC* (which increases the risk of colorectal cancer)[28] and factor V Leiden (which increases the risk of thrombosis).[29]

An example of the relative contributions of rare, highly penetrant mutations as opposed to common, less penetrant ones is seen in Alzheimer's disease. Rare mutations in presenilin 1, presenilin 2, or the β-amyloid precursor protein gene are highly penetrant causes of early-onset Alzheimer's disease; indeed, Alzheimer's disease develops by the age of 60 years in most people who are heterozygous for a mutation in one of these genes.[30] However, because so few people carry a mutation in any of these genes, these mutations play a part in fewer than 1 percent of cases of Alzheimer's disease.[31] In contrast, the apolipoprotein E ϵ4 allele also increases the risk of late-onset Alzheimer's disease (and atherosclerosis[32]), but more subtly. One representative Finnish study found that Alzheimer's disease develops during the mid-70s in approximately 8 percent of persons who are heterozygous for the ϵ4 allele and 21 percent of those who are homozygous for it, as compared with 3 percent of those with no ϵ4 allele.[33] Nonetheless, because approximately 26 percent of the U.S. population is heterozygous and 2 percent is homozygous[34] for the apolipoprotein E ϵ4 allele, this genetic factor has a role in many more cases of Alzheimer's disease than do the mutations in the genes for presenilin 1, presenilin 2, and β-amyloid precursor protein combined.

One characteristic of the human genome with medical and social relevance is that, on average, two unrelated persons share over 99.9 percent of their DNA sequences.[3] However, given the more than 3 billion base pairs that constitute the human genome, this also means that the DNA sequences of two unrelated humans vary at millions of bases. Since a person's genotype represents the blending of parental genotypes, we are each thus heterozygous at about 3 million bases. Many efforts are currently under way, in both the academic and commercial sectors, to catalogue these variants, commonly referred to as "single-nucleotide polymorphisms" (SNPs), and to correlate these specific genotypic variations with specific phenotypic variations relevant to health.

Some SNP–phenotype correlations occur as a direct result of the influence of the SNP on health. More commonly, however, the SNP is merely a marker of biologic diversity that happens to correlate with health because of its proximity to the genetic factor that is actually the cause. In this sense, the term "proximity" is only a rough measure of physical closeness; instead, it connotes that, as genetic material has passed through 5000 generations from our common African ancestral pool, recombination between the SNP and the actual genetic factor has occurred only rarely. In genetic terms, the SNP and the actual genetic factor are said to be in linkage disequilibrium (Fig. 2.4).

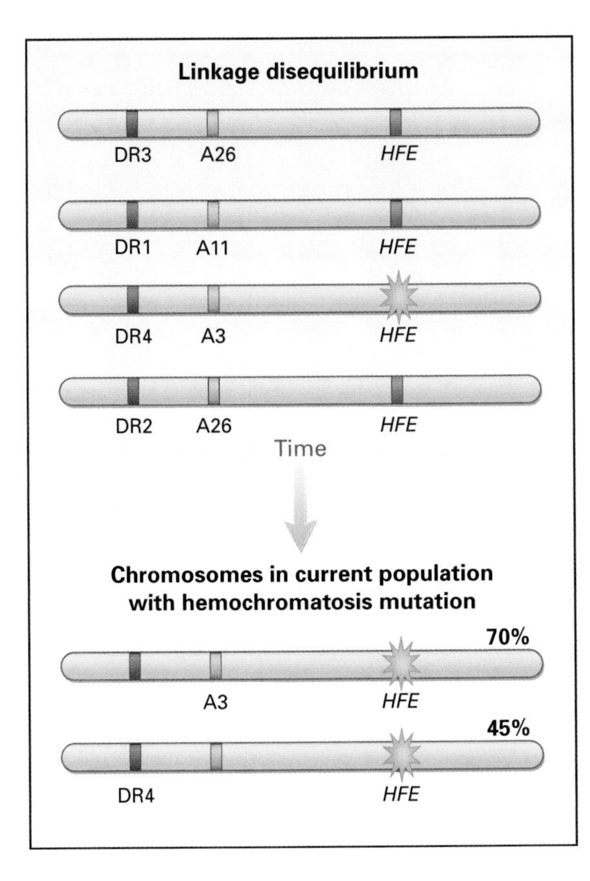

Figure 2.4 Example of Linkage Disequilibrium. As shown in the top panel, several versions of chromosome 6 existed in a specific population a number of generations ago. A mutation (indicated by the green star) in the hemochromatosis gene (*HFE*) originates in an ancestral chromosome that also carried the HLA-A3 and DR4 alleles. For several subsequent generations, nearly all chromosomes carrying the *HFE* mutation also had the HLA-A3 and DR4 alleles. As shown in the bottom panel, over time, recombination between the *HFE* and HLA-A alleles occurred rarely, so that in the current population, the *HFE* mutation is associated with the HLA-A3 allele 70 percent of the time, even though the HLA-A3 allele occurs on only 15 percent of normal chromosomes in this population. Because the HLA-DR locus is farther away from the *HFE* locus than is the HLA-A locus, recombination between the *HFE* and HLA-DR loci has occurred more frequently than between the *HFE* and the HLA-A loci, so that the *HFE* mutation is now associated with the HLA-DR4 allele 45 percent of the time, even though the HLA-DR4 allele occurs on only 25 percent of normal chromosomes in this population. The *HFE* mutation is said to be in strong linkage disequilibrium with HLA-A3 and somewhat weaker linkage disequilibrium with HLA-DR4.

An extension of the current efforts to catalogue SNPs and correlate them to phenotype are efforts to map and use haplotypes. Whereas a SNP represents a single-nucleotide variant, a haplotype represents a considerably longer sequence of nucleotides (averaging about 25,000), as well as any variants, that tend to be inherited together. SNPs and haplotypes will be the key to the association studies (i.e., studies of affected persons and control subjects) necessary to identify the genetic factors in complex, common diseases, just as family studies have been important to the identification of the genes involved in monogenic conditions.[35,36] Also, at least until whole-genome sequencing of individual patients becomes feasible clinically, the identification of SNPs and haplotypes will prove instrumental in efforts to use genomic medicine to individualize health care.

CONCLUSIONS

Except for monozygotic twins, each person's genome is unique. All physicians will soon need to understand the concept of genetic variability, its interactions with the environment, and its implications for patient care. With the sequencing of the human genome only months from its finish, the practice of medicine has now entered an era in which the individual patient's genome will help determine the optimal approach to care, whether it is preventive, diagnostic, or therapeutic. Genomics, which has quickly emerged as the central basic science of biomedical research, is poised to take center stage in clinical medicine as well.

References

1. Adams FL. The genuine works of Hippocrates. Vol. 2. New York: William Wood, 1886:338.

2. Lander ES, Linton LM, Birren B, et al. Initial sequencing and analysis of the human genome. Nature 2001; 409:860–921. [Errata, Nature 2001; 411: 720, 412:565.]

3. Venter JC, Adams MD, Myers EW, et al. The sequence of the human genome. Science 2001; 291: 1304–51. [Erratum, Science 2001; 292:1838.]

4. McKusick VA, Ruddle FH. A new discipline, a new name, a new journal. Genomics 1987; 1:1–2.

5. Armstrong K, Eisen A, Weber B. Assessing the risk of breast cancer. N Engl J Med 2000; 342:564–71.

6. Lynch HT. Hereditary nonpolyposis colorectal cancer (HNPCC). Cytogenet Cell Genet 1999; 86: 130–5.

7. Michael NL. Host genetic influences on HIV-1 pathogenesis. Curr Opin Immunol 1999; 11:466–74.

8. Small PM, Fujiwara PI. Management of tuberculosis in the United States. N Engl J Med 2001; 345: 189–200.

9. Mouradian MM. Recent advances in the genetics and pathogenesis of Parkinson disease. Neurology 2002; 58:179–85.

10. St George-Hyslop PH. Molecular genetics of Alzheimer disease. Semin Neurol 1999; 19:371–83.

11. McKusick VA. Mendelian inheritance in man: catalogs of human genes and genetic disorders. 12th ed. Baltimore: Johns Hopkins University Press, 1998.

12. Graveley BR. Alternative splicing: increased diversity in the proteomic world. Trends Genet 2001; 17:100–7.

13. Wallace DC. Mitochondrial diseases in man and mouse. Science 1999; 283:1482–8.

14. Online Mendelian Inheritance in Man, OMIM (TM). Baltimore: McKusick-Nathans Institute for Genetic Medicine, Johns Hopkins University, 2000. (Accessed October 15, 2002, at http://www.ncbi.nlm.nih.gov/omim/.)

15. Reik W, Walter J. Genomic imprinting: parental influence on the genome. Nat Rev Genet 2001; 2:21–32.

16. Robinson WP. Mechanisms leading to uniparental disomy and their clinical consequences. Bioessays 2000; 22:452–9.

17. Sinden RR. Biological implications of the DNA structures associated with disease-causing triplet repeats. Am J Hum Genet 1999; 64:346–53.

18. Goldstein JL, Hobbs HH, Brown MS. Familial hypercholesterolemia. In: Scriver CR, Beaudet AL, Sly WS, Valle D, eds. The metabolic & molecular bases of inherited disease. 8th ed. Vol. 2. New York: McGraw-Hill, 2001:2863–913.

19. Luzzatto L, Mehta A, Vulliamy T. Glucose 6-phosphate dehydrogenase deficiency. In: Scriver CR, Beaudet AL, Sly WS, Valle D, eds. The metabolic & molecular bases of inherited disease. 8th ed. Vol. 3. New York: McGraw-Hill, 2001:4517–53.

20. Goldberg YP, Nicholson DW, Rasper DM, et al. Cleavage of huntingtin by apopain, a proapoptotic cysteine protease, is modulated by the polyglutamine tract. Nat Genet 1996; 13:442–9.

21. Jin P, Warren ST. Understanding the molecular basis of fragile X syndrome. Hum Mol Genet 2000; 9:901–8.

22. Todd JA. From genome to aetiology in a multifactorial disease, type 1 diabetes. Bioessays 1999; 21: 164–74.

23. Grant SFA, Reid DM, Blake G, Herd R, Fogelman I, Ralston SH. Reduced bone density and osteoporosis associated with a polymorphic Sp1 binding site in the collagen type I α 1 gene. Nat Genet 1996; 14:203–5.

24. Horikawa Y, Oda N, Cox NJ, et al. Genetic variation in the gene encoding calpain-10 is associated with type 2 diabetes mellitus. Nat Genet 2000; 26:163–75. [Erratum, Nat Genet 2000; 26:502.]

25. Dean M, Carrington M, Winkler C, et al. Genetic restriction of HIV-1 infection and progression to AIDS by a deletion allele of the *CKR5* structural gene. Science 1996; 273:1856–62. [Erratum, 1996; 274:1069.]

26. Nathanson KN, Wooster R, Weber BL. Breast cancer genetics: what we know and what we need. Nat Med 2001; 7:552–6. [Erratum, Nat Med 2001; 7:749.]

27. Froguel P, Velho G. Molecular genetics of maturity-onset diabetes of the young. Trends Endocrinol Metab 1999; 10:142–6.

28. Fearnhead NS, Britton MP, Bodmer WF. The ABC of APC. Hum Mol Genet 2001; 10:721–33.

29. Major DA, Sane DC, Herrington DM. Cardiovascular implications of the factor V Leiden mutation. Am Heart J 2000; 140:189–95.

30. Campion D, Dumanchin C, Hannequin D, et al. Early-onset autosomal dominant Alzheimer disease: prevalence, genetic heterogeneity, and mutation spectrum. Am J Hum Genet 1999; 65:664–70.

31. Roses AD. Alzheimer diseases: a model of gene mutations and susceptibility polymorphisms for complex psychiatric diseases. Am J Med Genet 1998; 81:49–57.

32. Curtiss LK, Boisvert WA. Apolipoprotein E and atherosclerosis. Curr Opin Lipidol 2000; 11:243–51.

33. Kuusisto J, Koivisto K, Kervinen K, et al. Association of apolipoprotein E phenotypes with late onset Alzheimer's disease: population based study. BMJ 1994; 309:636–8.

34. Roses AD. Apolipoprotein E alleles as risk factors in Alzheimer's disease. Annu Rev Med 1996; 47: 387–400.

35. Reich DE, Cargill M, Bolk S, et al. Linkage disequilibrium in the human genome. Nature 2001; 411:199–204.

36. Goldstein DB, Weale ME. Population genetics: linkage disequilibrium holds the key. Curr Biol 2001; 11:R576–R579.

3

Genetic Testing

WYLIE BURKE, M.D., Ph.D.

Genetic testing can provide dramatic clinical benefits. A child known to have multiple endocrine neoplasia type 2 (MEN-2) can be spared medullary carcinoma by undergoing prophylactic thyroidectomy (Fig. 3.1),[1] and an adult with hereditary hemochromatosis can be spared cirrhosis by the early initiation of phlebotomy treatment.[2] Genetic testing can also provide diagnostic and prognostic information that aids in difficult clinical decision making. For example, a test for a deletion in the dystrophin gene, the cause of Duchenne's muscular dystrophy, can be used to identify women who are carriers of this condition (Fig. 3.2).[3] A carrier may avoid having an affected child by avoiding pregnancy or by undergoing prenatal testing for Duchenne's muscular dystrophy, with possible pregnancy termination if the fetus is found to be affected.

As these examples illustrate, most available genetic tests address questions related to rare or uncommon diseases. Even hemochromatosis, often described as a common genetic disease, has a prevalence of 0.5 percent or less.[4] However, the scope of genetic testing is expanding to include tests that assess the genetic risk of common diseases such as cancer and cardiovascular disease.[5,6]

DEFINITION OF GENETIC TESTING

A genetic test is "the analysis of human DNA, RNA, chromosomes, proteins, and certain metabolites in order to detect heritable disease-related genotypes, mutations, phenotypes, or karyotypes for clinical purposes."[7] This definition reflects the broad range of techniques that can be used in the testing process. Genetic tests also have diverse purposes, including the diagnosis of genetic disease in newborns, children, and adults; the identification of future health risks; the prediction of drug responses; and the assessment of risks to future children. Examples of currently available genetic tests are given in Tables 3.1 and

Originally published December 5, 2002.

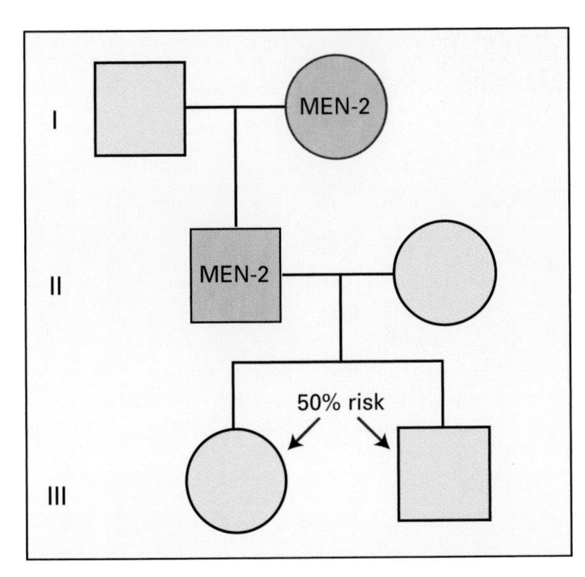

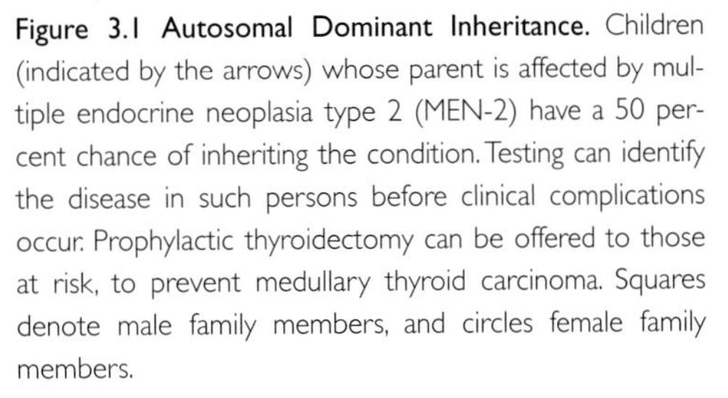

Figure 3.1 Autosomal Dominant Inheritance. Children (indicated by the arrows) whose parent is affected by multiple endocrine neoplasia type 2 (MEN-2) have a 50 percent chance of inheriting the condition. Testing can identify the disease in such persons before clinical complications occur. Prophylactic thyroidectomy can be offered to those at risk, to prevent medullary thyroid carcinoma. Squares denote male family members, and circles female family members.

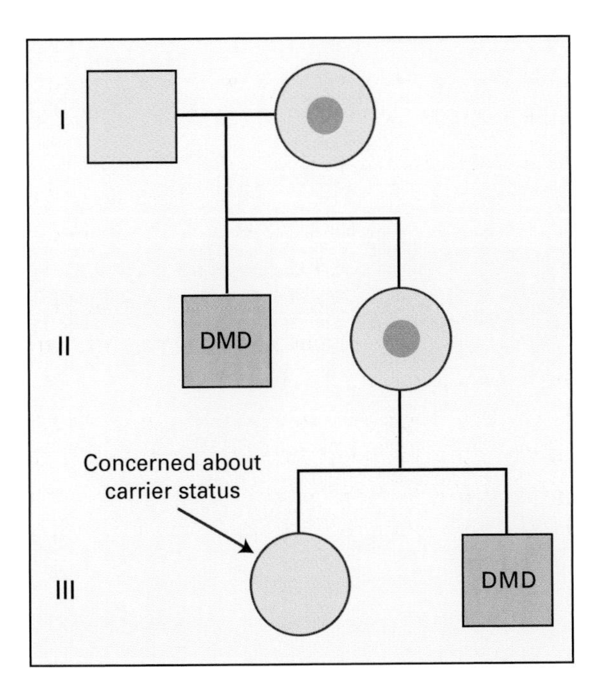

Figure 3.2 X-Linked Recessive Inheritance. A woman (indicated by the arrow) wants to know whether she carries the gene for Duchenne's muscular dystrophy (DMD), because her uncle and her brother were both affected (blue symbols), and her mother and grandmother are known to be carriers (symbols with an orange center). She has a 50 percent chance of inheriting the carrier status from her mother. Genetic testing can be used to determine her carrier status if her affected brother has a positive test result. Squares denote male family members, and circles female family members.

3.2,[8–11] and a comprehensive and continually updated listing of available tests can be found at the GeneTests–GeneClinics Web site (http://www.geneclinics.org).[8]

GENETIC DIAGNOSIS

Genetic testing is often the best way to confirm a diagnosis in a patient with signs or symptoms suggestive of a genetic disease. The technique chosen depends on both the clinical question and the predictive value of the available tests. For a young patient with medullary cancer of the thyroid, for example, the identification of a mutation in the *RET* oncogene

Table 3.1 Examples of Genetic Tests*

Condition and Available Tests	Comment
Multiple endocrine neoplasia type 2 Molecular tests Panel test: DNA-based detection of common *RET* gene mutations Sequencing: analysis of DNA sequence of specific coding regions of the gene to detect sequence variation	An autosomal dominant condition causing a high lifetime risk of MTC. The disease has 3 subtypes: 2A, associated with onset of MTC in childhood or early adulthood and an increased risk of parathyroid adenoma or hyperplasia; familial medullary thyroid carcinoma associated with a risk of MTC alone, with onset usually in adulthood; and 2B, associated with onset of MTC in early childhood and with characteristic facial features, mucosal neuromas of the lips and tongue, and ganglioneuromatosis of the gastrointestinal tract.
Duchenne's muscular dystrophy Molecular tests: DNA-based detection of deletions or structural inversions in coding regions of the Duchenne's muscular dystrophy gene Biochemical test: measurement of dystrophin protein in muscle tissue by weight (values are 0–3 percent of normal values) or by immunohistochemical techniques (showing complete or nearly complete absence of dystrophin)	An X-linked recessive condition causing progressive skeletal-muscle weakness and cardiomyopathy. Affected children are typically wheelchair-bound by the age of 12 years. Death is usually due to cardiomyopathy or respiratory failure.
Sickle cell anemia Molecular test: DNA-based detection of the HbS mutation Hematologic test: hemoglobin electrophoresis detects HbS	An autosomal recessive condition causing vasoocclusive events, resulting in pain crises, cerebrovascular complications, and splenic and renal dysfunction. Sickle cell anemia results from the HbS/HbS genotype. Related sickle cell disorders are caused by HbS in combination with other β-globin variants, such as HbC.
Cystic fibrosis Molecular tests Panel test: DNA-based detection of common *CFTR* mutations; core panel recommended by the American College of Obstetrics and Gynecology and the American College of Medical Genetics contains 25 mutations and is estimated to identify 85 percent of carriers in the general North American population Biochemical test: detection of elevated sweat chloride concentration	An autosomal recessive condition causing progressive lung disease. Most affected patients also have pancreatic insufficiency; other common complications include chronic sinusitis, meconium ileus, and male infertility.
Down's syndrome Chromosome test: staining of chromosomes to detect extra copy of chromosome 21	A chromosomal condition causing mental retardation and characteristic facial features. Other complications may include congenital heart disease and childhood leukemia.
22q11 deletion syndrome Chromosome test: fluorescence in situ hybridization to detect small deletions in chromosome 22	A chromosomal microdeletion causing learning difficulties, congenital heart disease, palatal abnormalities, and characteristic facial features.

(*continued*)

Table 3.1 (*continued*)

Condition and Available Tests	Comment
Iron overload Molecular tests: DNA-based detection of C282Y and H63D mutations in the *HFE* gene Biochemical test: measurement of transferrin saturation (serum iron ÷ TIBC × 100): if elevated (>60 for men, >50 for women), serum ferritin measured; if elevated (>300 μg/liter for men, >200 μg/liter for women), iron status assessed by liver biopsy or serial phlebotomy	A condition causing excess iron accumulation, resulting in deposition of iron in body tissues. Complications include cirrhosis, primary liver cancer, cardiomyopathy, diabetes, joint pain, and impotence. Most primary iron overload in the United States is due to mutations in the *HFE* gene, but other genetic disorders of iron overload have also been described.
Venous thromboembolism Molecular test for factor V Leiden: DNA-based detection of the factor V Leiden mutation Biochemical test for factor V Leiden: measurement of activated protein C resistance	The most common genetic risk factor is factor V Leiden, a mutation in the factor V gene. Other genetic contributors include the prothrombin variant 20210A, antithrombin III deficiency, protein S deficiency, and protein C deficiency.
Breast and ovarian cancer Molecular tests Panel test: DNA-based detection of 2 *BRCA1* mutations and 1 *BRCA2* mutation common in Ashkenazi Jewish populations Sequencing: analysis of DNA sequence of coding regions and adjacent segments of *BRCA1* and *BRCA2* to detect sequence variation	Mutations in the *BRCA1* and *BRCA2* genes are associated with an increased risk of breast and ovarian cancer.

*Additional information about these conditions and available genetic tests can be found at http://www.geneclinics.org,[8] http://www.cdc.gov/genomics/hugenet/reviews.htm,[9] http://www.cancer.gov/cancer_information/pdq,[10] and http://www.ncbi.nlm.nih.gov/omim/.[11] MTC denotes medullary carcinoma of the thyroid, HbS hemoglobin S, HbC hemoglobin C, and TIBC total iron-binding capacity.

confirms that the cancer is a manifestation of MEN-2, which accounts for approximately one quarter of cases of medullary thyroid cancer. The *RET*-mutation test can identify 85 to 95 percent of affected relatives of patients with medullary carcinoma.[12–14]

Testing for dystrophin gene deletions with the use of DNA-based technology is now the preferred diagnostic test for Duchenne's muscular dystrophy when clinical signs and symptoms suggest the diagnosis. A positive test confirms the diagnosis. A muscle biopsy is needed if the DNA-based test is negative. A negative test occurs in about 30 percent of patients with Duchenne's muscular dystrophy because some mutations in the dystrophin gene are not detected by current DNA testing.[15]

This level of genetic complexity is common and is termed "allelic heterogeneity," meaning that there are multiple different mutations (or alleles) in the same gene, all of which may lead to disease. For example, hundreds of different disease-causing mutations have been found in the cystic fibrosis gene[16] and the *BRCA1* and *BRCA2* genes associated with susceptibility to breast and ovarian cancer.[17]

In contrast, the most common form of sickle cell anemia, a disease occurring in 1 in 700 blacks in the United States, is caused by a single specific mutation in the β-globin gene,

Table 3.2 Examples of Molecular Genetic Tests*

Condition	Genes	Reported Uses of Testing
Neurologic		
Spinocerebellar ataxias	*SCA1, SCA2, SCA3, SCA6, SCA7, SCA10, DRPLA*	Diagnostic, predictive
Early-onset familial Alzheimer's disease	*PSEN1, PSEN2*	Diagnostic, predictive
Canavan's disease	*ASPA*	Diagnostic, prenatal
Nonsyndromic inherited congenital hearing loss (without other medical complications)	*GJB2*	Diagnostic, prenatal
Fragile X syndrome	*FMR1*	Diagnostic, prenatal
Huntington's disease	*HD*	Diagnostic, predictive, prenatal
Connective tissue		
Ehlers–Danlos syndrome, vascular type	*COL3A1*	Diagnostic, prenatal
Marfan's syndrome	*FBN1*	Diagnostic, prenatal
Osteogenesis imperfecta types I–IV	*COL1A1, COL1A2*	Diagnostic, prenatal
Oncologic		
Familial adenomatous polyposis	*APC*	Diagnostic, predictive
Hereditary nonpolyposis colorectal cancer	*MLH1, MSH2, PMS2, MSH3, MSH6*	Diagnostic, predictive
von Hippel–Lindau disease	*VHL*	Diagnostic, predictive
Li–Fraumeni syndrome	*TP53*	Diagnostic, predictive
Hematologic		
β-thalassemia	β-globin *(HbB)*	Carrier detection, prenatal diagnosis
Hemophilia A	*F8C*	Prognostic, carrier detection, prenatal
Hemophilia B	*F9C*	Carrier detection, prenatal
Renal		
Nephrogenic diabetes insipidus	*AVPR2, AQP2*	Diagnostic, carrier detection, prenatal
Polycystic kidney disease (autosomal dominant and autosomal recessive)	*PKD1, PKD2, PKHD1*	Predictive, prenatal
Multisystem		
Achondroplasia	*FGFR3*	Prenatal
Alpha$_1$-antitrypsin deficiency†	*AAT*	Diagnostic, predictive
Cystinosis	*CTNS*	Carrier detection, prenatal
Galactosemia	*GALT*	Newborn screening, carrier detection, prenatal‡
Neurofibromatosis type 1	*NF1*	Prenatal
Neurofibromatosis type 2	*NF2*	Predictive, prenatal

*This table is intended to be illustrative, not exhaustive. Most entries are based on information from GeneTest–GeneClinics at http://www.geneclinics.org; this Web site includes a comprehensive list of available molecular genetic tests and further clinical information about these and other genetic conditions.

†Biochemical testing is done to identify the enzyme deficiency.

‡Newborn screening is done by biochemical testing.

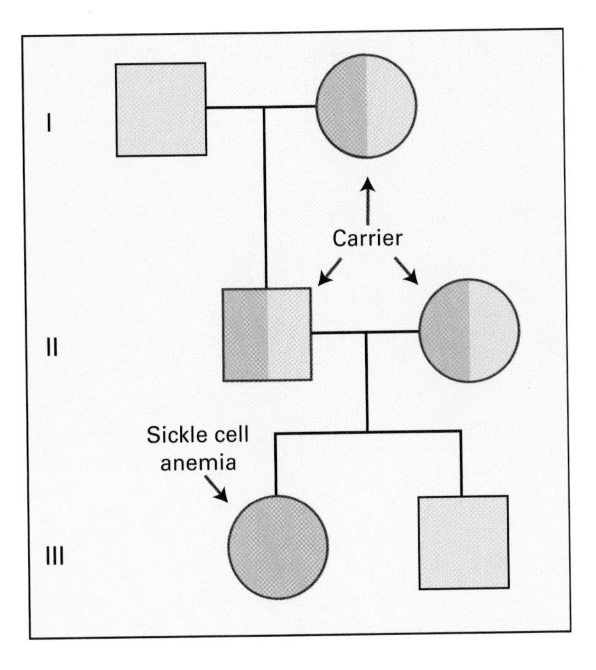

Figure 3.3 Autosomal Recessive Inheritance. When an autosomal recessive condition such as sickle cell anemia is diagnosed in a child (indicated by the arrow), the parents are identified as carriers of the sickle cell trait, which is inherited. All children of these parents have a 25 percent chance of being affected. Children who do not have sickle cell anemia have a 67 percent chance of being carriers. Cystic fibrosis is also inherited as an autosomal recessive condition.

resulting in a modified hemoglobin, termed hemoglobin S (HbS).[18] Both hematologic and DNA-based tests are available. Diagnostic testing can be done reliably by hemoglobin electrophoresis, but the DNA-based test for the HbS mutation is an important additional option because it makes prenatal diagnosis possible (Fig. 3.3).[19]

Cytogenetic tests are used to diagnose chromosomal disorders, in which chromosomes or chromosomal segments are duplicated, deleted, or translocated to different chromosomes. These tests make it possible to identify the chromosomal basis of conditions such as Down's syndrome (Table 3.1), which are caused by the presence of an extra chromosome, the lack of a chromosomal segment, or rearrangement of the chromosomes.[20,21] One cytogenetic technique, fluorescence in situ hybridization, identifies specific chromosomal regions through the use of fluorescent DNA probes and thus can pinpoint small chromosomal duplications and deletions missed by previous methods.[22,23] For example, the 22q11 deletion syndrome, a genetic condition caused by small deletions of chromosome 22 (Fig. 3.4), is characterized by a variety of learning disabilities, palatal abnormalities, and congenital heart disease.[24] Using fluorescence in situ hybridization, it has been possible to show that six previously described clinical syndromes, each with an overlapping cluster of physical and cognitive deficits, all represent manifestations of the 22q11 deletion syndrome.[24]

FAMILIAL RISK

A genetic diagnosis often indicates that other family members are at risk for the same condition. Genetic testing can help in evaluating this risk. For example, when the causative mutation of a genetic condition is known, presymptomatic diagnosis of family members is often possible and may offer an important opportunity for disease prevention. Thus, after a person

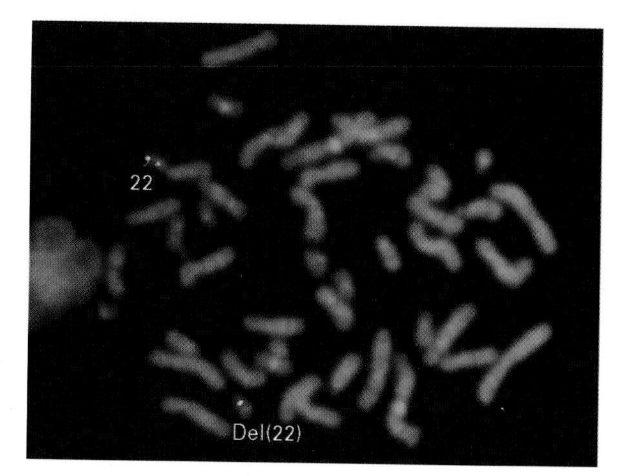

Figure 3.4 Fluorescence in Situ Hybridization Showing the 22q11 Microdeletion Syndrome. An orange probe identifies the chromosomal segment that is deleted in the syndrome; thus, the chromosome 22 with the microdeletion—del(22)—lacks this probe. A green probe identifies a different segment of the chromosome and is used as a marker for the two copies of chromosome 22, one of which is normal and thus demonstrates both probes (22). Photomicrograph provided courtesy of Dr. Christine Disteche and Douglas Chapman, University of Washington.

is given a diagnosis of MEN-2 and the causative *RET* mutation is identified, testing of all first-degree relatives is recommended (Fig. 3.1) so that prophylactic thyroidectomy can be offered to those who inherited the mutation.[25,26] A small number of other inherited cancer syndromes, such as familial adenomatous polyposis, offer a similar opportunity.[27]

The identification of risk does not necessarily lead to treatment options, however. Genetic testing for Huntington's disease, an autosomal dominant condition that causes progressive motor and cognitive dysfunction starting in midlife, allows people with an affected parent to determine whether they have inherited the causative mutation.[28] If the mutation is present, the person's risk of Huntington's disease is virtually 100 percent, given a normal life span. Yet, no effective intervention or preventive treatment is currently available. The choice to be tested is thus highly personal, and test results have the potential to be stigmatizing or psychologically harmful. For this reason, careful pretest counseling is recommended. A 10-year experience in the United Kingdom suggests that only about 20 percent of those at risk for Huntington's disease pursue such testing.[28]

In the case of X-linked and autosomal recessive conditions (Fig. 3.2 and 3.3), the purpose of genetic testing is often to identify family members who are carriers—that is, persons who are themselves unaffected but who are at risk of having affected children. As with decisions about testing for Huntington's disease, tests to determine carrier status are done primarily for personal, rather than medical, reasons: in this case to facilitate decisions about having children. For women who are carriers of an X-linked recessive disease, each son has a 50 percent risk of inheriting the disease (Fig. 3.2). With autosomal recessive diseases, such as sickle cell anemia or cystic fibrosis (Fig. 3.3), the risk of having an affected child is incurred only if both parents are carriers and is 25 percent for each pregnancy. If carrier status is confirmed, prenatal testing can be offered to provide an opportunity to inform parents about the genetic diagnosis before the birth, so that they can decide what course of action is best for them.

Prenatal diagnosis is also commonly used to diagnose Down's syndrome. This genetic condition is rarely inherited; most cases are due to an error in the formation of ovum or sperm, leading to the inclusion of an extra chromosome 21 at conception.[29] As with prenatal diagnosis for inherited genetic diseases, this use of genetic testing is focused on reproductive decision making rather than on clinical management of genetic disease.

Genetic testing is also sometimes used to identify family members with mild cases. For example, mild cases of 22q11 deletion syndrome have been documented among parents and siblings of patients with the condition.[30] Identifying these affected relatives may explain otherwise unexpected clinical findings, and also provides information about recurrence risks within the family: if a parent is affected, the condition can be passed on to future children.

CLINICAL VALIDITY OF GENETIC TESTS

These different examples help illustrate the importance of a test's clinical validity, defined as the accuracy with which a test predicts a clinical outcome.[7] Clinical validity reflects both the sensitivity of the test—the proportion of affected people with a positive test—and the penetrance of the mutations identified by the test. Penetrance refers to the proportion of people with the mutation who will manifest the disease; in the case of genetic diseases like Duchenne's muscular dystrophy, the proportion is virtually 100 percent, whereas in the case of hereditary nonpolyposis colon cancer, an inherited colorectal cancer syndrome, about 75 percent are likely to be affected.

Many DNA-based tests have reduced sensitivity because they identify only a subgroup of potentially causative mutations. This limitation is due to the state of scientific knowledge—some causative mutations may not yet be known—and to the properties of clinically available tests. For some conditions, a test for all known mutations would be prohibitively expensive, leading to a pragmatic tradeoff between cost and sensitivity. Just as scientific knowledge and costs change over time, so will the sensitivity and predictive value of various tests.

Reduced sensitivity has important implications for the testing of family members. For example, when a child with Duchenne's muscular dystrophy is found to have a deletion involving the dystrophin gene, the carrier status of female relatives can be determined by the same test. However, if the affected child does not have an identifiable mutation, the test cannot be used effectively either to determine carrier status or for prenatal diagnosis. An alternative approach—linkage analysis—is possible if two or more family members are affected and available for testing; this approach identifies patterns of DNA markers associated with the disease in a particular family (Fig. 3.5). But if the affected child is the only known member of the family with Duchenne's muscular dystrophy, linkage cannot be established, and this approach will not work.

When a genetic test has high sensitivity, people can be tested for carrier status without reference to the test results of an affected family member. This is the case for sickle cell anemia, which is caused by a specific mutation in the β-globin gene (Fig. 3.3).[18] In contrast, testing for cystic fibrosis can identify many (but not all) carriers in the general population; currently available tests identify the most common mutations and in the process usually identify 85 percent of carriers in the U.S. population.[16,31] (The use of genetic testing in population screening is discussed in chapter 4.)

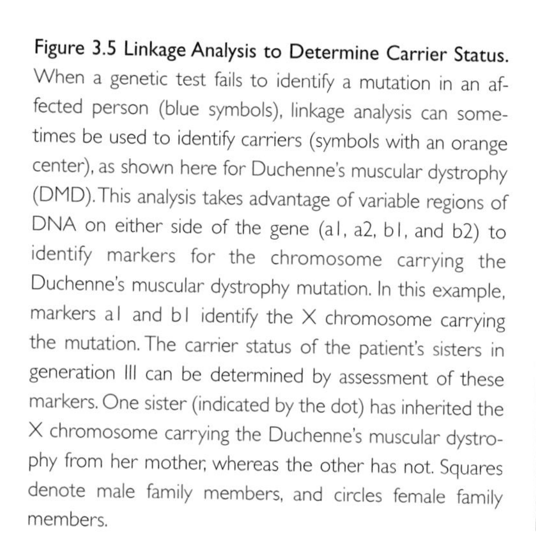

Figure 3.5 Linkage Analysis to Determine Carrier Status. When a genetic test fails to identify a mutation in an affected person (blue symbols), linkage analysis can sometimes be used to identify carriers (symbols with an orange center), as shown here for Duchenne's muscular dystrophy (DMD). This analysis takes advantage of variable regions of DNA on either side of the gene (a1, a2, b1, and b2) to identify markers for the chromosome carrying the Duchenne's muscular dystrophy mutation. In this example, markers a1 and b1 identify the X chromosome carrying the mutation. The carrier status of the patient's sisters in generation III can be determined by assessment of these markers. One sister (indicated by the dot) has inherited the X chromosome carrying the Duchenne's muscular dystrophy from her mother, whereas the other has not. Squares denote male family members, and circles female family members.

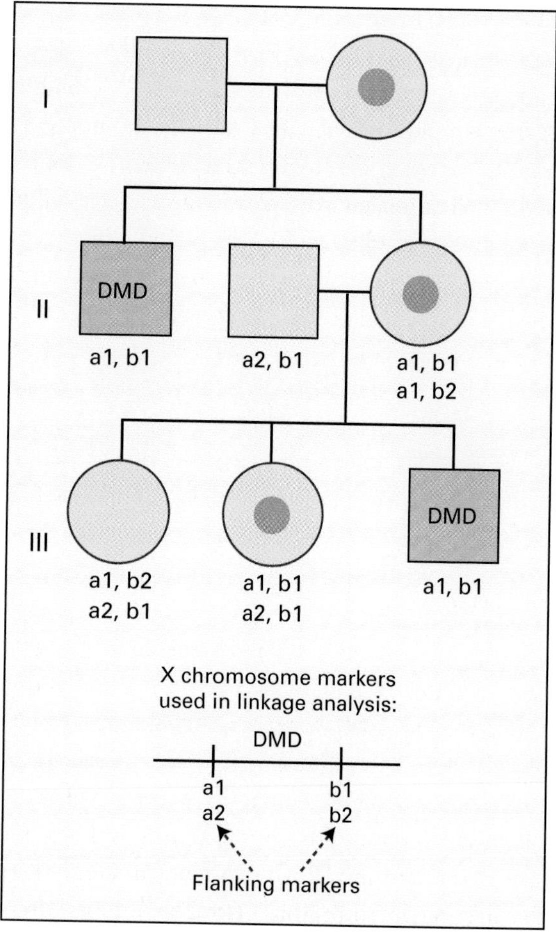

Most well-defined genetic diseases are caused by mutations with a high rate of penetrance and, as a result, have a high positive predictive value—that is, the likelihood of disease is high when the test is positive. This observation may contribute to the perception that current genetic tests are always highly predictive. However, even when mutations are highly penetrant, the negative predictive value of a test—the likelihood that disease is absent if the test is negative—can be low, if the test fails to identify all causative mutations.

GENETIC TESTING TO IMPROVE PREVENTIVE CARE

Genetic tests can also be used to determine genetic contributions to the risk of common diseases, in order to guide preventive care. Testing for *BRCA1* and *BRCA2* mutations provides an opportunity to identify people who may benefit from tailored screening and prevention protocols that are based on their genetic susceptibility to breast and ovarian cancer.[32–34] Estimates of the lifetime risk of breast cancer associated with these mutations range from 26 to 85 percent; the risk of ovarian cancer is also elevated but to a lesser extent, and risk estimates also vary.[35–41]

In conditions with a low rate of penetrance, more evidence is needed to establish the efficacy of interventions to reduce risk.[42,43] In the case of MEN-2, the evidence favoring prophylactic thyroidectomy derives from the observation of a low rate of medullary thyroid cancer among patients who had the surgery.[25,26] The power of such studies derives from historical data demonstrating a lifetime risk of cancer of close to 100 percent in patients with untreated MEN-2, with an associated high rate of premature mortality.[1] When a genetic test predicts an increased risk rather than a certainty of future disease, the efficacy of interventions to reduce risk is more difficult to measure,[42] particularly when the level of risk is uncertain, as is the case with *BRCA1* and *BRCA2*. If the risk is initially overestimated—a common bias when mutations conferring risk are found in families selected for high risk—the efficacy of an intervention may be greatly overestimated in the absence of controlled observations.[38]

This issue will take on greater importance as genetic factors conferring smaller risks are identified.[44,45] Mutations associated with a high risk account for only a small percentage of common diseases; mutations in *BRCA1* and *BRCA2* are a rare cause of breast cancer, for example. The largest genetic contribution to health is in the form of common variants that increase or decrease risk to a moderate degree.[5,46,47] These tests have lower positive and negative predictive values than most currently available genetic tests, but they have potential implications for a larger number of people and are an important byproduct of the Human Genome Project.[5] Two examples offer insights into the implications of genetic tests of this kind: hemochromatosis and factor V Leiden.

Hemochromatosis, a condition involving excess accumulation of iron, can lead to iron overload, which in turn can result in complications such as cirrhosis, diabetes, cardiomyopathy, and arthritis.[4] Two mutations in the *HFE* gene, C282Y and H63D, promote excess accumulation of iron. C282Y is the more severe mutation, and the C282Y/C282Y genotype accounts for the majority of clinically penetrant cases.[4] But current data suggest that clinical disease does not develop in a substantial proportion of people with this genotype.[48,49] A pooled analysis found that patients with the *HFE* genotypes C282Y/H63D and H63D/H63D are also at increased risk for iron overload,[50] yet overall, disease is likely to develop in fewer than 1 percent of people with these genotypes. Thus, DNA-based tests for hemochromatosis identify a genetic risk rather than the disease itself.[51] Environmental factors such as diet and exposure to alcohol or other hepatotoxins may modify the clinical outcome in patients with hemochromatosis,[4] and variations in other genes affecting iron metabolism may also be a factor.[52] As a result, the clinical condition of iron overload is most reliably diagnosed on the basis of biochemical evidence of excess body iron.[2,4] Whether it is beneficial to screen asymptomatic people for a genetic risk of iron overload is a matter of debate.[47,53]

Factor V Leiden offers another example. This factor V gene mutation is relatively common, ranging in prevalence from 1 to 5 percent in different American ethnic groups,[54] and results in up to an eightfold increased risk of venous thrombosis.[55,56] Estimates of the annual incidence of venous thrombosis in people who are heterozygous for factor V Leiden range from 0.19 to 0.58 percent,[57–59] suggesting a lifetime risk of 12 to 30 percent. However, more than half of the thromboembolic events associated with factor V Leiden occur when other risk factors, such as surgery, use of oral contraceptives, and bed rest, are also present.[55,57,60] Both gene–gene and gene–environment interactions contribute to the overall risk of venous thrombosis.[55] Thus, factor V Leiden, like mutations in the *HFE* gene, is a risk factor for disease rather than an indication of the presence of disease.

As is the case for predictive testing for hemochromatosis, the clinical usefulness of testing for factor V Leiden is not established. Although a positive test identifies people at increased risk for venous thrombosis, the implications for management are unclear. Interventions such as prophylactic anticoagulation therapy or avoidance of risk factors might be considered, but evidence of the clinical benefit of such interventions is so far lacking.[61,62]

The issue of specificity of treatment is an important one. New genetic tests to assess the risk of common diseases are likely to have properties similar to those of tests for factor V Leiden. They will identify relatively common genetic traits that interact with other genetic and environmental factors to increase risk. Their clinical usefulness will depend on the availability of specific, effective interventions to reduce risk. In the absence of genotype-specific interventions, the knowledge of a person's genetic susceptibility to a condition could result in worry or job- or insurance-related discrimination without yielding health benefits or could even be harmful to a person's health by reducing motivation to pursue risk-reducing measures.[63]

INFORMED CONSENT AND GENETIC COUNSELING

Patients are usually given detailed counseling before undergoing genetic testing, to ensure that they make informed decisions about the use of tests with complex personal implications. Genetic counseling is traditionally "nondirective"—that is, counseling provides sufficient information to allow families or individual persons to determine the best course of action for themselves but avoids making testing recommendations.[64,65] This approach was developed in the context of genetic tests for reproductive decision making and untreatable conditions such as Huntington's disease, in which the value of testing is based on personal preference.

When tests are conducted to improve clinical management, pretest counseling needs differ. A testing recommendation is appropriate, for example, when a test offers an opportunity to prevent disease, as in the case with testing for MEN-2.[66] Since some tests for genetic risk factors will probably become a routine part of clinical practice, they are likely to be offered without formal pretest counseling. In approaching the question of informed consent to conduct a specific genetic test, however, the potential social and family implications need to be acknowledged, including the potential for discrimination on the basis of genetic-risk status[7] and the possibility that the predictive value of genetic information may be overestimated.[67] These considerations suggest that clinicians should err on the side of caution and follow carefully-thought-out informed-consent procedures for genetic testing, unless outcome studies suggest otherwise.

CONCLUSIONS

Genetic testing offers important opportunities for diagnosis and assessment of genetic risk. The sensitivity of tests for rare conditions will continue to improve as additional causative mutations are identified. Genetic tests are available to determine the risk of common diseases, but these often have limited predictive value. Evaluating the clinical usefulness of these tests will require a careful assessment of the risks and benefits of testing; the availability of specific measures to reduce risk in genetically susceptible people will be a major consideration.

One of the difficult challenges in the use of genetic tests is a constantly changing knowledge base. Fortunately, a growing number of Internet sites are available to provide clinicians with up-to-date information (Table 3.1).[68,69] Research to evaluate interventions based on genetic risk will assume increasing importance as new tests become available. Because the development of tests to assess risk is likely to outpace the ability to reduce the risk, an ongoing dialogue involving clinicians and policymakers will be needed to develop a consensus about their appropriate clinical use.

Acknowledgments

Supported in part by a grant (R01-HG02263) from the National Institutes of Health. The contents of this article are solely the responsibility of the author and do not necessarily represent the official views of the National Institutes of Health.

References

1. Hoff AO, Cote GJ, Gagel RF. Multiple endocrine neoplasias. Annu Rev Physiol 2000;62: 377–411.

2. Bacon BR, Sadiq SA. Hereditary hemochromatosis: presentation and diagnosis in the 1990s. Am J Gastroenterol 1997; 92:784–9.

3. Abbs S, Bobrow M. Report on the 16th ENMC workshop—carrier diagnosis of Duchenne and Becker muscular dystrophy. Neuromuscul Disord 1993; 3:241–2.

4. Hanson EH, Imperatore G, Burke W. HFE gene and hereditary hemochromatosis: a HuGE review. Am J Epidemiol 2001; 154:193–206.

5. Collins FS. Shattuck Lecture—medical and societal consequences of the Human Genome Project. N Engl J Med 1999; 341:28–37.

6. Roses AD. Pharmacogenetics and the practice of medicine. Nature 2000; 405:857–65.

7. Holtzman NA, Watson MS, eds. Promoting safe and effective genetic testing in the United States: final report of the Task Force on Genetic Testing. Baltimore: Johns Hopkins University Press, 1999.

8. GeneTests–GeneClinics home page. Seattle: University of Washington, 2002. (Accessed November 8, 2002, at http://www.geneclinics.org.)

9. The Human Genome Epidemiology Network: HuGE reviews. Atlanta: Centers for Disease Control and Prevention, 2002. (Accessed October 8, 2002, at http://www.cdc.gov/genomics/hugenet/reviews.htm.)

10. CancerNet PDQ. Bethesda, Md.: National Cancer Institute, 2002. (Accessed November 8, 2002, at http://www.cancer.gov/cancer_information/pdq.)

11. OMIM: online Mendelian inheritance in man. Bethesda, Md.: National Center for Biotechnology Information, 2002. (Accessed October 8, 2002, at http://www.ncbi.nlm.nih.gov/omim/.)

12. Weisner GL, Snow K. Multiple endocrine neoplasia type 2 [includes: MEN 2A (Sipple syndrome), MEN 2B (mucosal neuroma syndrome), familial medullary thyroid carcinoma (FMTC)]. Seattle: GeneClinics, 1999. (Accessed November 8, 2002, at http://www.geneclinics.org/profiles/men2/details.html.)

13. Raue F. German medullary thyroid carcinoma/multiple endocrine neoplasia registry. Arch Surg 1998; 383:334–6.

14. Kebebew E, Ituarte PH, Siperstein AE, Duh QY, Clark OH. Medullary thyroid carcinoma: clinical characteristics, treatment, prognostic factors, and a comparison of staging systems. Cancer 2000; 88: 1139–48.

15. Korf BR, Darras BT, Urion DK. Dystrophinopathies [includes: Duchenne muscular dystrophy (DMD, pseudohypertrophic muscular dystrophy), Becker muscular dystrophy (BMD), and X-linked dilated cardiomyopathy (XLDCM)]. Seattle: GeneClinics, 2000. (Accessed November 8, 2002, at http://www.geneclinics.org/profiles/dbmd/details.html.)

16. Grody WW. Cystic fibrosis: molecular diagnosis, population screening, and public policy. Arch Pathol Lab Med 1999; 123:1041–6.

17. Brody LC, Biesecker BB. Breast cancer susceptibility genes: BRCA1 and BRCA2. Medicine (Baltimore) 1998; 77:208–26.

18. Ashley-Koch A, Yang Q, Olney RS. Sickle hemoglobin (HbS) allele and sickle cell disease: a HuGE review. Am J Epidemiol 2000; 151:839–45.

19. Cao A, Galanello R, Rosatelli MC. Prenatal diagnosis and screening of the haemoglobinopathies. Baillieres Clin Haematol 1998; 11:215–38.

20. Crow JF. Two centuries of genetics: a view from halftime. Annu Rev Genomics Hum Genet 2000; 1:21–40.

21. Capone GT. Down syndrome: advances in molecular biology and the neurosciences. J Dev Behav Pediatr 2001; 22:40–59.

22. Pergament E. New molecular techniques for chromosome analysis. Baillieres Best Pract Res Clin Obstet Gynaecol 2000; 14:677–90.

23. McDonald-McGinn D, Emanuel BS, Zackai EH, Children's Hospital of Philadelphia. 22q11 Deletion syndrome. [Includes: Shprintzen syndrome, DiGeorge syndrome (DGS), velocardiofacial syndrome (VCFS), conotruncal anomaly face syndrome (CTAF), Caylor cardiofacial syndrome, Opitz G/BBB]. Seattle: GeneClinics, 1999. (Accessed November 8, 2002, at http://www.geneclinics.org/profiles/22q11deletion/index.html.)

24. De Decker HP, Lawrenson JB. The 22q11.2 deletion: from diversity to a single gene theory. Genet Med 2001; 3:2–5.

25. Wells SA Jr, Skinner MA. Prophylactic thyroidectomy, based on direct genetic testing, in patients at risk for the multiple endocrine neoplasia type 2 syndromes. Exp Clin Endocrinol Diabetes 1998; 106:29–34.

26. Niccoli-Sire P, Murat A, Baudin E, et al. Early or prophylactic thyroidectomy in MEN 2/FMTC gene carriers: results in 71 thyroidectomized patients. Eur J Endocrinol 1999; 141:468–74.

27. Statement of the American Society of Clinical Oncology: genetic testing for cancer susceptibility, adopted on February 20, 1996. J Clin Oncol 1996; 14:1730–40.

28. Harper PS, Lim C, Craufurd D. Ten years of presymptomatic testing for Huntington's disease: the experience of the UK Huntington's Disease Prediction Consortium. J Med Genet 2000; 37:567–71.

29. Wald NJ, Hackshaw AK. Advances in antenatal screening for Down syndrome. Baillieres Best Pract Res Clin Obstet Gynaecol 2000; 14:563–80.

30. McDonald-McGinn DM, Tonnesen MK, Laufer-Cahana A, et al. Phenotype of the 22q11.2 deletion in individuals identified through an affected relative: cast a wide FISHing net! Genet Med 2001; 3:23–9.

31. Tait JF, Gibson RL, Marshall SG, Sternen DL, Cheng E, Cutting GR. Cystic fibrosis [CF, Mucovisiodosis: includes: congenital bilateral absence of the vas deferens (CBAVD)]. Seattle: GeneClinics, 2001. (Accessed November 8, 2002, at http://www.geneclinics.org/profiles/cf/details.html.)

32. Burke W, Daly M, Garber J, et al. Recommendations for follow-up care of individuals with an inherited predisposition to cancer. II. BRCA1 and BRCA2. JAMA 1997; 277:997–1003.

33. Hartmann LC, Schaid DJ, Woods JE, et al. Efficacy of bilateral prophylactic mastectomy in women with a family history of breast cancer. N Engl J Med 1999; 340:77–84.

34. CancerNet. CancerNet PDQ summary on breast/ovarian and colorectal cancer genetics. Bethesda, Md.: National Cancer Institute, 2001. (Accessed November 8, 2002, at http://cancer.gov/cancerinfo/pdq/genetics.)

35. Struewing JP, Hartge P, Wacholder S, et al. The risk of cancer associated with specific mutations of *BRCA1* and *BRCA2* among Ashkenazi Jews. N Engl J Med 1997; 336:1401–8.

36. Thorlacius S, Struewing JP, Hartge P, et al. Population-based study of risk of breast cancer in carriers of BRCA2 mutation. Lancet 1998; 352:1337–9.

37. Ford D, Easton DF, Stratton M, et al. Genetic heterogeneity and penetrance analysis of the BRCA1 and BRCA2 genes in breast cancer families. Am J Hum Genet 1998; 62:676–89.

38. Hopper JL, Southey MC, Dite GS, et al. Population-based estimate of the average age-specific cumulative risk of breast cancer for a defined set of protein-truncating mutations in BRCA1 and BRCA2: Australian Breast Cancer Family Study. Cancer Epidemiol Biomarkers Prev 1999; 8:741–7.

39. Warner E, Foulkes W, Goodwin P, et al. Prevalence and penetrance of BRCA1 and BRCA2 gene mutations in unselected Ashkenazi Jewish women with breast cancer. J Natl Cancer Inst 1999; 91:1241–7.

40. Prevalence and penetrance of BRCA1 and BRCA2 mutations in a population-based series of breast cancer cases. Br J Cancer 2000; 83:1301–8.

41. Satagopan JM, Offit K, Foulkes W, et al. The lifetime risks of breast cancer in Ashkenazi Jewish carriers of BRCA1 and BRCA2 mutations. Cancer Epidemiol Biomarkers Prev 2001; 10:467–73.

42. Welch HG, Burke W. Uncertainties in genetic testing for chronic disease. JAMA 1998; 280:1525–7.

43. Preventive Services Task Force. Guide to clinical preventive services: report of the U.S. Preventive Services Task Force. 2nd ed. Baltimore: Williams & Wilkins, 1996.

44. Ziv E, Cauley J, Morin PA, Saiz R, Browner WS. Association between the T29→C polymorphism in the transforming growth factor beta1 gene and breast cancer among elderly white women: the Study of Osteoporotic Fractures. JAMA 2001; 285:2859–63. [Erratum, JAMA 2001; 286:3081.]

45. Armstrong K. Genetic susceptibility to breast cancer: from the roll of the dice to the hand women were dealt. JAMA 2001; 285:2907–9.

46. Holtzman NA, Marteau TM. Will genetics revolutionize medicine? N Engl J Med 2000; 343:141–4.

47. Kaprio J. Science, medicine, and the future: genetic epidemiology. BMJ 2000; 320:1257–9.

48. Beutler E, Felitti VJ, Koziol JA, Ho NJ, Gelbart T. Penetrance of 845G→A (C282Y) HFE hereditary haemochromatosis mutation in the USA. Lancet 2002; 359:211–8.

49. Asberg A, Hveem K, Thorstensen K, et al. Screening for hemochromatosis: high prevalence and low morbidity in an unselected population of 65,238 persons. Scand J Gastroenterol 2001; 36:1108–15.

50. Burke W, Imperatore G, McDonnell SM, Baron RC, Khoury MJ. Contribution of different HFE genotypes to iron overload disease: a pooled analysis. Genet Med 2000; 2:271–7.

51. Adams P, Brissot P, Powell LW. EASL International Consensus Conference on Haemochromatosis. J Hepatol 2000; 33:485–504.

52. Andrews NC. Disorders of iron metabolism. N Engl J Med 1999; 341:1986–95. [Erratum, N Engl J Med 2000; 342:364.]

53. Cogswell ME, McDonnell SM, Khoury MJ, Franks AL, Burke W, Brittenham G. Iron overload, public health, and genetics: evaluating the evidence for hemochromatosis screening. Ann Intern Med 1998; 129:971–9.

54. Ridker PM, Miletich JP, Hennekens CH, Buring JE. Ethnic distribution of factor V Leiden in 4047 men and women: implications for venous thromboembolism screening. JAMA 1997; 277:1305–7.

55. Rosendaal F. Venous thrombosis: a multicausal disease. Lancet 1999; 353:1167–73.

56. Meyer G, Emmerich J, Helley D, et al. Factors V Leiden and II 20210A in patients with symptomatic pulmonary embolism and deep vein thrombosis. Am J Med 2001; 110:12–5.

57. Middeldorp S, Meinardi JR, Koopman MM, et al. A prospective study of asymptomatic carriers of the factor V Leiden mutation to determine the incidence of venous thromboembolism. Ann Intern Med 2001; 135:322–7.

58. Simioni P, Sanson BJ, Prandoni P, et al. Incidence of venous thromboembolism in families with inherited thrombophilia. Thromb Haemost 1999; 81:198–202.

59. Martinelli I, Bucciarelli P, Margaglione M, De Stefano V, Castaman G, Mannucci PM. The risk of venous thromboembolism in family members with mutations in the genes of factor V or prothrombin or both. Br J Haematol 2000; 111:1223–9.

60. Bloemenkamp KW, Rosendaal FR, Helmerhorst FM, Buller HR, Vandenbroucke JP. Enhancement by factor V Leiden mutation of risk of deep-vein thrombosis associated with oral contraceptives containing a third-generation progestagen. Lancet 1995; 346:1593–6.

61. Grody WW, Griffin JH, Taylor AK, Korf BR, Heit JA. American College of Medical Genetics consensus statement on factor V Leiden mutation testing. Genet Med 2001; 3:139–48.

62. Bauer KA. The thrombophilias: well-defined risk factors with uncertain therapeutic implications. Ann Intern Med 2001; 135:367–73.

63. Marteau TM, Lerman C. Genetic risk and behavioural change. BMJ 2001; 322:1056–9.

64. Genetic counseling. Am J Hum Genet 1975; 27:240–2.

65. Michie S, Smith JA, Heaversedge J, Read S. Genetic counseling: clinical geneticists' views. J Genet Couns 1999; 8:275–87.

66. Cummings S. The genetic testing process: how much counseling is needed? J Clin Oncol 2000; 18:Suppl:60S–64S.

67. Hubbard R, Lewontin RC. Pitfalls of genetic testing. N Engl J Med 1996; 334:1192–4.

68. Pagon RA, Pinsky L, Beahler CC. Online medical genetics resources: a US perspective. BMJ 2001; 322:1035–7.

69. Stewart A, Haites N, Rose P. Online medical genetics resources: a UK perspective. BMJ 2001; 322:1037–9.

4

Population Screening in the Age of Genomic Medicine

MUIN J. KHOURY, M.D., Ph.D., LINDA L. McCABE, Ph.D., and EDWARD R. B. McCABE, M.D., Ph.D.

Physicians in the era of genomic medicine will have the opportunity to move from intense, crisis-driven intervention to predictive medicine. Over the next decade or two, it seems likely that we will screen entire populations or specific subgroups for genetic information in order to target interventions to individual patients that will improve their health and prevent disease. Until now, population screening involving genetics has focused on the identification of persons with certain mendelian disorders before the appearance of symptoms and thus on the prevention of illness[1] (e.g., screening of newborns for phenylketonuria), the testing of selected populations for carrier status, and the use of prenatal diagnosis to reduce the frequency of disease in subsequent generations (e.g., screening to identify carriers of Tay–Sachs disease among Ashkenazi Jews). But in the future, genetic information will increasingly be used in population screening to determine individual susceptibility to common disorders such as heart disease, diabetes, and cancer. Such screening will identify groups at risk so that primary-prevention efforts (e.g., diet and exercise) or secondary-prevention efforts (early detection or pharmacologic intervention) can be initiated. Such information could lead to the modification of screening recommendations, which are currently based on population averages (e.g., screening of people over 50 years of age for the early detection of colorectal cancer).[2]

In this review, we describe current and evolving principles of population screening in genetics. We also provide examples of issues related to screening in the era of genomic medicine.

Originally published January 2, 2003

PRINCIPLES OF POPULATION SCREENING

The principles of population screening developed in 1968 by Wilson and Jungner[3] form a basis for applying genetics in population screening. These principles emphasize the importance of a given condition to public health, the availability of an effective screening test, the availability of treatment to prevent disease during a latent period, and cost considerations. Wald outlined three elements of screening: the identification of persons likely to be at high risk for a specific disorder so that further testing can be done and preventive actions taken, outreach to populations that have not sought medical attention for the condition, and follow-up and intervention to benefit the screened persons.[4] Several groups have used these principles to develop policies regarding genetic testing in populations.[5] Screening of newborns, which has been carried out in the United States since the early 1960s, serves as a foundation for other types of genetic screening.[6,7]

NEWBORN SCREENING

Each state (and the District of Columbia) determines its own list of diseases and methods for the screening of newborns. Only phenylketonuria and hypothyroidism are screened for by all these jurisdictions.[7] Table 4.1 lists the disorders that are included in many state programs for newborn screening and gives one an idea of the diversity of techniques employed. The addition of a test or a method to a state's screening program depends on the efforts of advisory boards for newborn screening, political lobbying of legislatures, and the efforts of laboratory personnel for newborn screening. There has often been a lack of research to demonstrate the effectiveness of screening and treatment for a disorder, either before or after the disease is added to the newborn-screening program. The technological spectrum ranges from the original Guthrie bacterial inhibition assay, developed in the late 1950s,[8] to tandem mass spectrometry[9,10] and DNA analysis.[11–13] With the use of DNA testing of the blood blot obtained from the screening of a newborn, the state of Texas reduced the age at confirmation of the diagnosis of sickle cell disease from four months to two months.[14] Rapid diagnostic confirmation is imperative for the initiation of penicillin prophylaxis to prevent illness and death in patients with sickle cell disease.[15,16] The cost of this follow-up test is $10 or less for each positive sample from the original screening.[14]

Two-tiered testing is also used for congenital hypothyroidism, since patients with primary hypothyroidism have elevated levels of thyrotropin and low levels of thyroxine.[17,18] The two-tiered strategy provides better sensitivity and specificity than either test alone. However, the health care professional needs to use clinical judgment in addition to the results of newborn screening. If a patient with a negative newborn-screening test has symptoms of congenital hypothyroidism, clinical acumen should override the test result and specific diagnostic testing should be performed.[17] The results of screening tests are not infallible because of the possibility of biologic, clerical, and laboratory errors.[19–21]

Audiometry is used to screen newborns for hearing defects. The frequency of deafness in childhood is as high as 1 in 500.[22] These programs are based in hospitals and are therefore decentralized.[7] Mutations in the gene for connexin 26 account for 40 percent of all cases of

Table 4.1 Disorders Included in Newborn-Screening Programs

Disorder	Screening Method	States Offering Test	Treatment
Phenylketonuria	Guthrie bacterial inhibition assay Fluorescence assay Amino-acid analyzer Tandem mass spectrometry	All	Diet restricting phenylalanine
Congenital hypothyroidism	Measurement of thyroxine and thyrotropin	All	Oral levothyroxine
Hemoglobinopathies	Hemoglobin electrophoresis Isoelectric focusing High-performance liquid chromatography Follow-up DNA analysis	Most	Prophylactic antibiotics Immunization against *Diplococcus pneumoniae* and *Haemophilus influenzae*
Galactosemia	Beutler test Paigen test	Limited no.	Galactose-free diet
Maple syrup urine disease	Guthrie bacterial inhibition assay	Limited no.	Diet restricting intake of branched-chain amino acids
Homocystinuria	Guthrie bacterial inhibition assay	Limited no.	Vitamin B_{12} Diet restricting methionine and supplementing cystine
Biotinidase deficiency	Colorimetric assay	Limited no.	Oral biotin
Congenital adrenal hyperplasia	Radioimmunoassay Enzyme immunoassay	Limited no.	Glucocorticoids Mineralocorticoids Salt
Cystic fibrosis	Immunoreactive trypsinogen assay followed by DNA testing Sweat chloride test	Limited no.	Improved nutrition Management of pulmonary symptoms

childhood hearing loss, with a carrier rate of 3 percent in the population.[23] A single mutation is responsible for most of these cases in a mixed U.S. population.[23] A different mutation is predominant among Ashkenazi Jews.[24] Two-tiered testing in which audiometry is followed by DNA testing for mutations in the connexin 26 gene may be a useful and cost-effective approach to screening for hearing loss.[25] Early detection provides the possibility of aggressive intervention to improve a child's language skills, provide cochlear implants, or do both.[23]

In 1999, the American Academy of Pediatrics and the Health Resources and Services Administration convened the Newborn Screening Task Force to address the lack of consistency in the disorders included in screening programs and the testing methods used in the various states.[26] The group concluded that there should be a national consensus on the diseases tested for in state programs of newborn screening. The American Academy of Pediatrics, American College of Medical Genetics, Health Resources and Services Administration, Centers for Disease Control and Prevention, March of Dimes, and other groups are working together to create a national agenda for newborn screening.

A disorder that may be included in newborn screening tests is cystic fibrosis. Cystic fibrosis has been included in the newborn-screening program in Colorado since the demonstration that some affected infants had malnutrition as a result of the pancreatic dysfunction.[27] This observation was confirmed by a randomized trial in Wisconsin involving infants with a positive newborn-screening test for cystic fibrosis.[28] In the study, infants with a positive test were randomly assigned to a screened group (in which physicians were informed of the positive screening result) or a control group (in which physicians were informed of the positive screening result when the child was four years of age if cystic fibrosis had not been diagnosed clinically or if the child's parents had not asked about the results of the screening test). In Wisconsin, infants are first tested with the use of an immunoreactive trypsinogen assay[29]; if the result is positive, the test is followed up with a DNA test of the original specimen of dried blood obtained for newborn screening.[30–32] The cost of each follow-up DNA test for infants with positive results on the immunoreactive trypsinogen assay was estimated to be $3 to $5.[31]

A new form of technology, tandem mass spectrometry, detects more than 20 disorders, not all of which can be treated. A justification for introducing tandem mass spectrometry is the identification of newborns with medium-chain acyl–coenzyme A (CoA) dehydrogenase deficiency (Fig. 4.1). Without early detection and intervention, this deficiency leads to episodic hypoglycemia, seizures, coma associated with intercurrent illnesses and fasting, and a risk of death of approximately 20 percent after the first episode in the first and second year of life.[33,34] Management of medium-chain acyl–CoA dehydrogenase deficiency involves educating families about the dangers of hypoglycemia, which can be triggered by fasting, with resulting fat catabolism, during intercurrent illnesses and by inadequate caloric intake, and of the need for aggressive intervention with intravenous glucose if hypoglycemia does occur. For many of the other disorders detected by tandem mass spectrometry, treatment is not available, but families will potentially be spared "diagnostic odysseys" with a severely ill child.[35] The eventual goal is collaborative research to determine the appropriate treatment after early diagnosis.[7,36] In addition, this information may be useful for genetic counseling of these families. A cause for concern is that tandem mass spectrometry may detect metabolic variations of unknown clinical significance, creating unwarranted anxiety in parents and health care professionals.

CARRIER SCREENING OF ADULT POPULATIONS FOR SINGLE-GENE DISORDERS

Tay–Sachs Disease

Carrier screening for Tay–Sachs disease has targeted Ashkenazi Jewish populations of childbearing age.[37] In a 30-year period, 51,000 carriers have been identified, resulting in the identification of 1400 two-carrier couples.[37] Another approach has been taken in

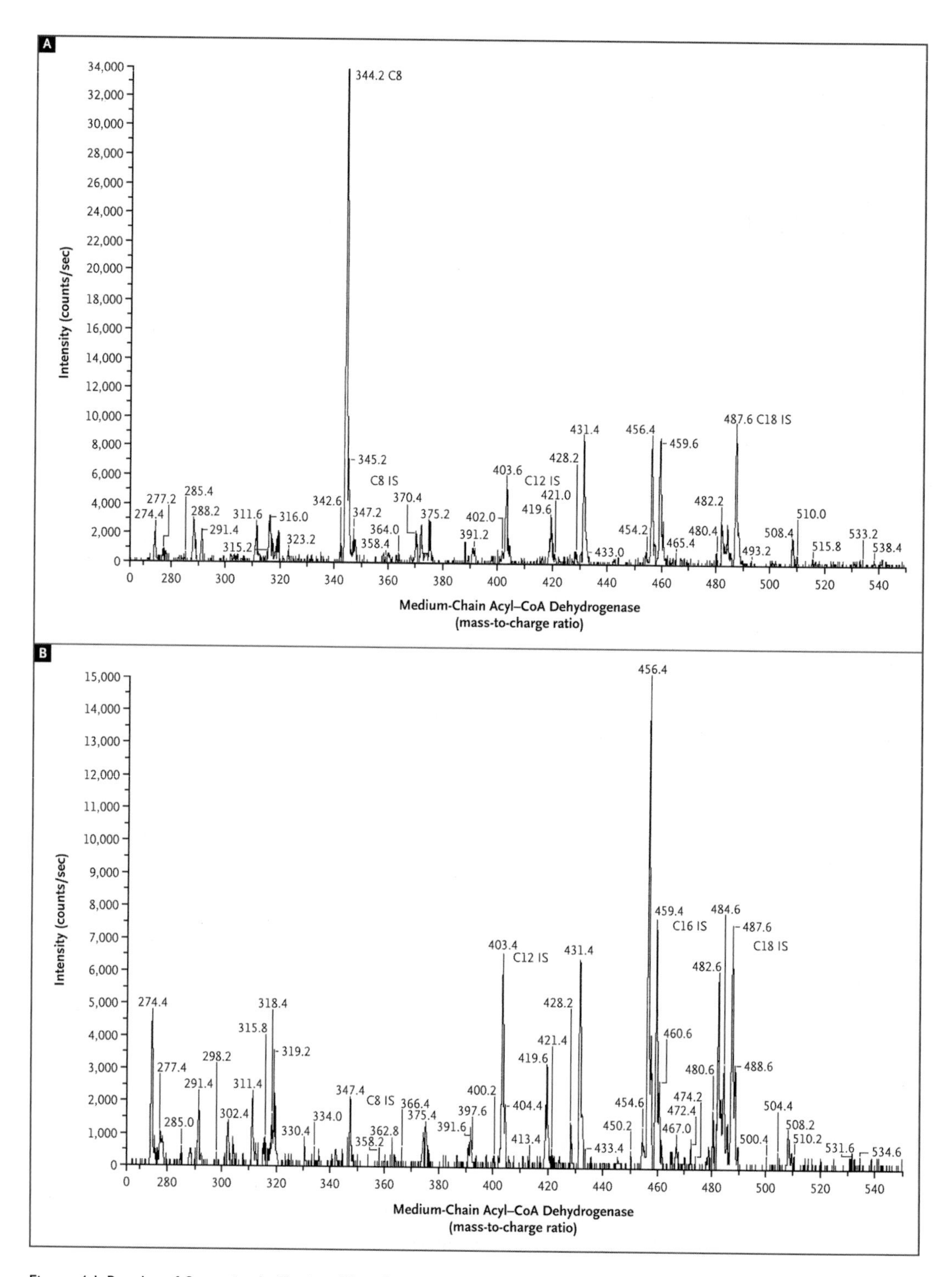

Figure 4.1 Results of Screening by Tandem Mass Spectrometry for Medium-Chain Acyl–Coenzyme A (CoA) Dehydrogenase Deficiency in an Affected Patient (Panel A) and a Control Subject (Panel B). IS denotes internal standard. Figure provided courtesy of John E. Sherwin, Ph.D.

Montreal, where high-school students learn about Tay–Sachs disease and thalassemia as part of a biology course. Those of Ashkenazi Jewish descent can request carrier testing for Tay–Sachs disease, and those of Mediterranean ancestry can be tested for thalassemia.[38] When women who have been identified as carriers in high school later consider becoming pregnant, they bring their partners in for testing. Although this program has been very successful in Canada, the culture and the legal environment in the United States, including a standard that does not allow high-school students to consent to medical care and the implications for insurability, may prohibit the adoption of such a model.[39]

Cystic Fibrosis

Northern Europeans have a carrier frequency of cystic fibrosis of 1 in 25 to 1 in 30; the rate is lower in other ethnic and cultural groups.[17] A 1997 National Institutes of Health Consensus Development Conference[40] recommended that the following populations be screened for mutations associated with cystic fibrosis: the adult family members of patients with cystic fibrosis, the partners of patients with cystic fibrosis, couples planning a pregnancy, and couples seeking prenatal care. Since more than 900 different mutations associated with cystic fibrosis have been reported in the literature,[41] the establishment of screening programs has been difficult. However, the American College of Medical Genetics, the American College of Obstetricians and Gynecologists, and the National Institutes of Health agreed that mutations with a carrier frequency of at least 0.1 percent in the general population should be screened for, resulting in a panel of 25 mutations recommended for carrier testing.[42] These guidelines suggest that carrier testing should be offered to all non-Jewish white persons and Ashkenazi Jews and that other ethnic and cultural groups should be informed of the limitations of the panel to detect carriers in their group (in the case of black persons) or of the low incidence of cystic fibrosis in their group (in the cases of Asian and Native American persons).

Mutations in the gene associated with cystic fibrosis have also been associated with obstructive azoospermia in men[43] and with chronic rhinosinusitis.[44,45] The guidelines recommend including in the screening panel a test for the R117H mutation, which is associated with congenital bilateral absence of the vas deferens.[42] If the R117H mutation is found, further testing and genetic counseling are recommended.[42]

POPULATION SCREENING FOR GENETIC SUSCEPTIBILITY TO COMMON DISEASES

Several groups have addressed the value of population screening for genetic susceptibility to conditions with onset in adulthood.[46–48] Table 4.2 presents a synthesis of the suggested modifications to the 1968 criteria,[3] based on current principles.

Hereditary hemochromatosis and the thrombophilia that results from carrying a single copy of a factor V Leiden gene are two adult-onset illnesses to which the suggested revised principles for population screening would apply (Table 4.2), and these illnesses also reflect the complex scientific and social issues involved in screening for risk factors for disease. As shown by Wald et al.,[49] screening for risk factors for nondiscrete traits that are

Table 4.2 Principles of Population Screening as Applied to Genetic Susceptibility to Disease*

Public health assessment

The disease or condition should be an important public health burden to the target population in terms of illness, disability, and death.

The prevalence of the genetic trait in the target population and the burden of disease attributable to it should be known.

The natural history of the condition, from susceptibility to latent disease to overt disease, should be adequately understood.

Evaluation of tests and interventions

Data should be available on the positive and negative predictive values of the test with respect to a disease or condition in the target population.

The safety and effectiveness of the test and accompanying interventions should be established.

Policy development and screening implementation

Consensus regarding the appropriateness of screening and interventions for people with positive and negative test results should be based on scientific evidence.

Screening should be acceptable to the target population.

Facilities should be available for adequate surveillance, prevention, treatment, education, counseling, and social support.

Screening should be a continual process, including pilot programs, evaluation of laboratory quality and health services, evaluation of the effect of screening, and provisions for changes on the basis of new evidence.

The cost effectiveness of screening should be established.

Screening and interventions should be accessible to the target population.

There should be safeguards to ensure that informed consent is obtained and the privacy of those tested is respected, that there is no coercion or manipulation, and that those tested are protected against stigmatization and discrimination.

*Principles are based on Wilson and Jungner,[3] Goel,[46] Khoury et al.,[47] and Burke et al.[48]

distributed continuously may not be beneficial even if the factors are associated with a high risk of disease (e.g., high cholesterol levels and heart disease). This is because risk factors are determined by comparing the probability of disease at each end of the distribution of the risk factor (those with the highest level of risk and those with the lowest level of risk). Those with a moderate level of risk are not considered. The likelihood of a disorder, given a positive screening result, is expressed relative to the average risk of the entire population. The goal of screening is to identify individual persons with a high risk in comparison to everyone else.

Hereditary Hemochromatosis

Many consider hereditary hemochromatosis to be the key example of the need for population screening in the genomic era,[50] but gaps in our knowledge preclude the recommendation of population screening for this disorder. This policy issue was discussed by an expert-panel workshop held by the Centers for Disease Control and Prevention and the National Human Genome Research Institute.[51] The panel concluded that population genetic testing for mutations in *HFE*, the gene for hereditary hemochromatosis, could not be recommended because of uncertainty about the natural history of the disease, age-related penetrance, optimal care for persons without symptoms who are found to carry mutations, and the psychosocial impact of genetic testing.[52,53] On the other hand, mutation analysis may be useful in confirming the diagnosis of hereditary hemochromatosis in persons with abnormal indexes of iron metabolism. A meta-analysis of studies[54] showed that homozygosity for the C282Y mutation was associated with the highest risk of hereditary hemochromatosis. The risks associated with other genotypes, including C282Y/H63D and H63D/H63D, were much lower. A recent large cohort study in the Kaiser Permanente Southern California health care network suggests that the disease penetrance for *HFE* mutations may be quite low.[55] Only 1 of the 152 subjects who were homozygous for C282Y had symptoms of hereditary hemochromatosis.

Several questions remain regarding the benefits and risks of identifying and treating persons without symptoms who are at high risk for hereditary hemochromatosis (i.e., through population screening). This process should be clearly distinguished from early case finding, which could include testing of iron status, and analysis for mutations in *HFE*, in persons who present with clinical symptoms consistent with a diagnosis of hereditary hemochromatosis. The natural history of hereditary hemochromatosis—particularly age-related penetrance—remains unknown. Despite the relatively high prevalence of the two most common mutations in the U.S. population,[56] questions persist regarding the nature and prevalence of mutations in specific ethnic and cultural groups, as well as the morbidity[57] and mortality[58] associated with this disease. Therefore, questions remain concerning the persons most likely to benefit from early treatment and thus about the optimal timing of screening and effective intervention, as well as ethical and psychosocial issues[59] (Table 4.2).

Factor V Leiden

Factor V is an important component of the coagulation cascade leading to the conversion of prothrombin into thrombin and the formation of clots.[60] In factor V Leiden, the triplet coding for arginine (CGA) at codon 506 is replaced by CAA, which codes for glutamine (R506Q), resulting in thrombophilia or an increased propensity for clot formation.[61] The prevalence of factor V Leiden varies.[62,63] Among persons of northern European descent, the prevalence is about 5 percent. The highest prevalence of factor V Leiden is found in Sweden and in some Middle Eastern countries; it is virtually absent in African and Asian populations. Heterozygosity for factor V Leiden results in an increase in the incidence of venous thrombosis by a factor of 4 to 9.[64,65]

An interaction between factor V Leiden and the use of oral contraceptives was originally found in a case–control study of risk factors for venous thrombosis.[66] Although the use of oral contraceptives alone increases the risk of venous thrombosis by a factor of about 4 and the presence of factor V Leiden alone increases the risk by a factor of about 7, their joint effect was an increase by a factor of more than 30. In spite of the high relative risk, the absolute risk was relatively low (about 28 per 10,000 person-years) among women with factor V Leiden who used oral contraceptives, because the incidence of this complication is relatively low in the population.

The question of whether it is beneficial to screen women for factor V Leiden before prescribing oral contraceptives remains controversial. Venous thrombosis is relatively rare, and the mortality associated with venous thrombosis is low among young women.[67] More than half a million women would need to be screened for factor V Leiden—resulting in tens of thousands of women being denied prescriptions for oral contraceptives—to prevent a single death. In addition to medical and financial considerations, there are issues related to the quality of life, the risk of illness and death from unwanted pregnancy, and concern about possible discrimination by insurance companies. In 2001, the American College of Medical Genetics stated that the opinions and practices regarding testing for factor V Leiden vary considerably, and no consensus has emerged.[68]

For the individual healthy woman contemplating the use of oral contraceptives, the risk–benefit equation does not currently favor screening. For women without symptoms who have family histories of multiple thrombosis, there are no evidence-based guidelines, and decisions will have to be reached individually, without reliance on population-based recommendations.

These examples show why it is essential that data continue to be analyzed to inform decision making for individual persons and populations.

ETHICAL, LEGAL, AND SOCIAL ISSUES

The following are among the ethical, legal, and social issues involved in population-based screening that confront health care providers, policymakers, and consumers.

Testing Children for Adult-Onset Disorders

Two committees of the American Academy of Pediatrics have recently addressed the issue of molecular genetic testing of children and adolescents for adult-onset disease.[69,70] The Committee on Genetics[69] recommended that persons under 18 years of age be tested only if testing offers immediate medical benefits or if another family member benefits and there is no anticipated harm to the person being tested. The committee regarded genetic counseling before and after testing as an essential part of the process.

The Committee on Bioethics[70] agreed with the Newborn Screening Task Force[27] that the inclusion of tests in the newborn-screening battery should be based on evidence and that there should be informed consent for newborn screening (which is currently not required in the majority of states). The Committee on Bioethics did not support the use of carrier screening in persons under 18 years of age, except in the case of an adolescent who is pregnant or is planning a pregnancy. It recommended against predictive testing for adult-onset disorders in persons under 18 years.

Unanticipated Information

Misattribution of Paternity

The American Society of Human Genetics has recommended that family members not be informed of misattributed paternity unless determination of paternity was the purpose of the test.[71] However, it must be recognized that such a policy may lead to misinformation regarding genetic risk.

Unexpected Associations among Diseases

In the course of screening for one disease, information regarding another disease may be discovered. Although the person may have requested screening for the first disorder, the presence of the second disorder may be unanticipated and may lead to stigmatization and discrimination on the part of insurance companies and employers. Informed consent should include cautions regarding unexpected findings from the testing.

Oversight and Policy Issues

In 1999, the Secretary's Advisory Committee on Genetic Testing was established to advise the Department of Health and Human Services on the medical, scientific, ethical, legal, and social issues raised by the development and use of genetic tests (http://www4.od.nih.gov/oba/sacgt.htm).[72] The committee conducted public outreach to identify issues regarding genetic testing. There was an overwhelming concern on the part of the public regarding discrimination in employment and insurance. The advisory committee recommended the support of legislation preventing discrimination on the basis of genetic information and increased oversight of genetic testing. The Food and Drug Administration was charged as the lead agency and was urged to take an innovative approach and consult experts outside the agency. The goal is to generate specific language for the labeling of genetic tests, much as drugs are described in the *Physicians' Desk Reference*.[73] Such labeling would provide persons considering, and health professionals recommending, genetic tests with information about the clinical validity and value of the test—what information the test will provide, what choices will be available to people after they know their test results, and the limits of the test.

In conclusion, although the use of genetic information for population screening has great potential, much careful research must be done to ensure that such screening tests, once introduced, will be beneficial and cost effective.

References

1. Juengst ET. "Prevention" and the goals of genetic medicine. Hum Gene Ther 1995; 6:1595–605.
2. Ransohoff DF, Sandler RS. Screening for colorectal cancer. N Engl J Med 2002; 346:40–4.
3. Wilson JMG, Jungner G. Principles and practice of screening for disease. Public health papers no. 34. Geneva: World Health Organization, 1968.
4. Wald NJ. The definition of screening. J Med Screen 2001; 8:1.
5. Wilfond BS, Thomson EJ. Models of public health genetics policy development. In: Khoury MJ, Burke W, Thomson EJ, eds. Genetics and public health in the 21st century: using genetic information to improve health and prevent disease. New York: Oxford University Press, 2000:61–82.
6. Committee for the Study of Inborn Errors of Metabolism. Genetic screening: programs, principles, and research. Washington, D.C.: National Academy of Sciences, 1975.
7. McCabe LL, Therrell BL Jr, McCabe ERB. Newborn screening: rationale for a comprehensive, fully integrated public health system. Mol Genet Metab 2002; 77:267–73.
8. Guthrie R, Susi A. A simple phenylalanine method for detecting phenylketonuria in large populations of newborn infants. Pediatrics 1963; 32:338–43.
9. Chace DH, Hillman SL, Van Hove JL, Naylor EW. Rapid diagnosis of MCAD deficiency: quantitative analysis of octanoylcarnitine and other acylcarnitines in newborn blood spots by tandem mass spectrometry. Clin Chem 1997; 43:2106–13.
10. Andresen BS, Dobrowolski SF, O'Reilly L, et al. Medium-chain acyl-CoA dehydrogenase (MCAD) mutations identified by MS/MS-based prospective

screening of newborns differ from those observed in patients with clinical symptoms: identification and characterization of a new, prevalent mutation that results in mild MCAD deficiency. Am J Hum Genet 2001; 68:1408–18.

11. McCabe ERB, Huang S-Z, Seltzer WK, Law ML. DNA microextraction from dried blood spots on filter paper blotters: potential applications to newborn screening. Hum Genet 1987; 75:213–6.

12. Jinks DC, Minter M, Tarver DA, Vanderford M, Hejtmancik JF, McCabe ERB. Molecular genetic diagnosis of sickle cell disease using dried blood specimens on blotters used for newborn screening. Hum Genet 1989; 81:363–6.

13. Descartes M, Huang Y, Zhang Y-H, et al. Genotypic confirmation from the original dried blood specimens in a neonatal hemoglobinopathy screening program. Pediatr Res 1992; 31:217–21.

14. Zhang Y-H, McCabe LL, Wilborn M, Therrell BL Jr, McCabe ERB. Application of molecular genetics in public health: improved follow-up in a neonatal hemoglobinopathy screening program. Biochem Med Metab Biol 1994; 52:27–35.

15. Gaston MH, Verter JI, Woods G, et al. Prophylaxis with oral penicillin in children with sickle cell anemia: a randomized trial. N Engl J Med 1986; 314:1593–9.

16. Consensus Development Panel. Newborn screening for sickle cell disease and other hemoglobinopathies. NIH consensus statement. Vol. 6. No. 9. Bethesda, Md.: NIH Office of Medical Applications of Research, 1987:1–22.

17. American Academy of Pediatrics Committee on Genetics. Newborn screening fact sheets. Pediatrics 1989; 83:449–64.

18. Burrow GN, Dussault JH, eds. Neonatal thyroid screening. New York: Raven Press, 1980:155.

19. McCabe ERB, McCabe L, Mosher GA, Allen RJ, Berman JL. Newborn screening for phenylketonuria: predictive validity as a function of age. Pediatrics 1983; 72:390–8.

20. Holtzman C, Slazyk WE, Cordero JF, Hannon WH. Descriptive epidemiology of missed cases of phenylketonuria and congenital hypothyroidism. Pediatrics 1986; 78:553–8.

21. Dequeker E, Cassiman J-J. Quality evaluation of data interpretation and reporting. Am J Hum Genet 2001; 69:Suppl:438. abstract.

22. Mehl AL, Thomson V. The Colorado Newborn Hearing Screening Project, 1992–1999: on the threshold of effective population-based universal newborn hearing screening. Pediatrics 2002; 109:134. abstract.

23. Cohn ES, Kelley PM. Clinical phenotype and mutations in connexin 26 (*DFNB1/GJB2*), the most common cause of childhood hearing loss. Am J Med Genet 1999; 89:130–6.

24. Morrell RJ, Kim HJ, Hood LJ, et al. Mutations in the connexin 26 gene (*GJB2*) among Ashkenazi Jews with nonsyndromic recessive deafness. N Engl J Med 1998; 339:1500–5.

25. McCabe ERB, McCabe LL. State-of-the-art for DNA technology in newborn screening. Acta Paediatr Suppl 1999; 88:58-60.

26. Newborn Screening Task Force. Serving the family from birth to the medical home: newborn screening: a blueprint for the future—a call for a national agenda on state newborn screening programs. Pediatrics 2000; 106:389–422.

27. Reardon MC, Hammond KB, Accurso FJ, et al. Nutritional deficits exist before 2 months of age in some infants with cystic fibrosis identified by screening test. J Pediatr 1984; 105:271–4.

28. Farrell PM, Kosorok MR, Rock MJ, et al. Early diagnosis of cystic fibrosis through neonatal screening prevents severe malnutrition and improves long-term growth. Pediatrics 2001; 107:1–13.

29. Hassemer DJ, Laessig RH, Hoffman GL, Farrell PM. Laboratory quality control issues related to screening newborns for cystic fibrosis using immunoreactive trypsin. Pediatr Pulmonol Suppl 1991; 7:76–83.

30. Seltzer WK, Accurso F, Fall MZ, et al. Screening for cystic fibrosis: feasibility of molecular genetic analysis of dried blood specimens. Biochem Med Metab Biol 1991; 46:105–9.

31. Gregg RG, Wilfond BS, Farrell PM, Laxova A, Hassemer D, Mischler EH. Application of DNA analysis in a population-screening program for neonatal diagnosis of cystic fibrosis (CF): comparison of screening protocols. Am J Hum Genet 1993; 52:616–26.

32. Kant JA, Mifflin TE, McGlennen R, Rice E, Naylor E, Cooper DL. Molecular diagnosis of cystic fibrosis. Clin Lab Med 1995; 15:877–98.

33. Roe CR, Ding J. Mitochondrial fatty acid oxidation disorders. In: Scriver CR, Beaudet AL, Sly WS, Valle D, eds. The metabolic & molecular bases of inherited disease. 8th ed. Vol. 2. New York: McGraw-Hill, 2001:2297–326.

34. Matsubara Y, Narisawa K, Tada K, et al. Prevalence of K329E mutation in medium-chain acyl-CoA dehydrogenase gene determined from Guthrie cards. Lancet 1991; 338:552–3.

35. Wilcken B, Travert G. Neonatal screening for cystic fibrosis: present and future. Acta Paediatr Suppl 1999; 88:33–5.

36. Naylor EW, Chace DH. Automated tandem mass spectrometry for mass newborn screening for

disorders in fatty acid, organic acid, and amino acid metabolism. J Child Neurol 1999; 14:Suppl 1:S4–S8.

37. Kaback MM. Population-based genetic screening for reproductive counseling: the Tay-Sachs disease model. Eur J Pediatr 2000; 159:Suppl 3:S192–S195.

38. Mitchell JJ, Capua A, Clow C, Scriver CR. Twenty-year outcome analysis of genetic screening programs for Tay-Sachs and β-thalassemia disease carriers in high schools. Am J Hum Genet 1996; 59:793–8.

39. McCabe L. Efficacy of a targeted genetic screening program for adolescents. Am J Hum Genet 1996; 59:762–3.

40. Genetic testing for cystic fibrosis: National Institutes of Health Consensus Development Conference statement on genetic testing for cystic fibrosis. Arch Intern Med 1999; 159:1529–39.

41. Grody WW, Desnick RJ. Cystic fibrosis population carrier screening: here at last—are we ready? Genet Med 2001; 3:87–90.

42. Grody WW, Cutting GR, Klinger KW, Richards CS, Watson MS, Desnick RJ. Laboratory standards and guidelines for population-based cystic fibrosis carrier screening. Genet Med 2001; 3:149–54.

43. Mak V, Zielenski J, Tsui L-C, et al. Proportion of cystic fibrosis gene mutations not detected by routine testing in men with obstructive azoospermia. JAMA 1999; 281:2217–24.

44. Raman V, Clary R, Siegrist KL, Zehnbauer B, Chatila TA. Increased prevalence of mutations in the cystic fibrosis transmembrane conductance regulator in children with chronic rhinosinusitis. Pediatrics 2002; 109:136–7. abstract.

45. Wang XJ, Moylan B, Leopold DA, et al. Mutation in the gene responsible for cystic fibrosis and predisposition to chronic rhinosinusitis in the general population. JAMA 2000; 284:1814–9.

46. Goel V. Appraising organised screening programmes for testing for genetic susceptibility to cancer. BMJ 2001; 322:1174–8.

47. Khoury MJ, Burke W, Thomson EJ. Genetics and public health: a framework for the integration of human genetics into public health practices. In: Khoury MJ, Burke W, Thomson EJ, eds. Genetics and public health in the 21st century: using genetic information to improve health and prevent disease. New York: Oxford University Press, 2000:3–24.

48. Burke W, Coughlin SS, Lee NC, Weed DL, Khoury MJ. Application of population screening principles to genetic screening for adult-onset conditions. Genet Test 2001; 5:201–11.

49. Wald NJ, Hackshaw AK, Frost CD. When can a risk factor be used as a worthwhile screening test? BMJ 1999; 319:1562–5.

50. Collins FS. Keynote speech at the Second National Conference on Genetics and Public Health, December 1999. Atlanta: Office of Genetics & Disease Prevention, 2000. (Accessed December 6, 2002, at http://www.cdc.gov/genomics/info/conference/intro.htm.)

51. Cogswell ME, Burke W, McDonnell SM, Franks AL. Screening for hemochromatosis: a public health perspective. Am J Prev Med 1999; 16:134–40.

52. Burke W, Thomson E, Khoury MJ, et al. Hereditary hemochromatosis: gene discovery and its implications for population-based screening. JAMA 1998; 280:172–8.

53. EASL International Consensus Conference on Hemochromatosis. III. Jury document. J Hepatol 2000; 33:496–504.

54. Burke W, Imperatore G, McDonnell SM, Baron RC, Khoury MJ. Contribution of different *HFE* genotypes to iron overload disease: a pooled analysis. Genet Med 2000; 2:271–7.

55. Beutler E, Felitti VJ, Koziol JA, Ho NJ, Gelbart T. Penetrance of 845G→A (C282Y) *HFE* hereditary haemochromatosis mutation in the USA. Lancet 2002; 359:211–8.

56. Steinberg KK, Cogswell ME, Chang JC, et al. Prevalence of C282Y and H63D mutations in the hemochromatosis (HFE) gene in the United States. JAMA 2001; 285:2216–22.

57. Brown AS, Gwinn M, Cogswell ME, Khoury MJ. Hemochromatosis-associated morbidity in the United States: an analysis of the National Hospital Discharge Survey, 1979–1997. Genet Med 2001; 3:109–11.

58. Yang Q, McDonnell SM, Khoury MJ, Cono J, Parrish RG. Hemochromatosis-associated mortality in the United States from 1979 to 1992: an analysis of Multiple-Cause Mortality Data. Ann Intern Med 1998; 129:946–53.

59. Imperatore G, Valdez R, Burke W. Case study: hereditary hemochromatosis. In: Khoury MJ, Little J, Burke W, eds. Human genome epidemiology: scientific foundation for using genetic information to improve health and prevent disease. New York: Oxford University Press, 2003:495–500.

60. Greenberg DL, Davie EW. Introduction to hemostasis and the vitamin K-dependent coagulation factors. In: Scriver CR, Beaudet AL, Sly WS, Valle D, eds. The metabolic & molecular bases of inherited disease. 8th ed. Vol. 3. New York: McGraw-Hill, 2001:4293–326.

61. Esmon CT. Anticoagulation protein C/thrombomodulin pathway. In: Scriver CR, Beaudet AL, Sly WS, Valle D, eds. The metabolic & molecular bases of inherited disease. 8th ed. Vol. 3. New York: McGraw-Hill, 2001:4327–43.

62. Rees DC, Cox M, Clegg JB. World distribution of factor V Leiden. Lancet 1995; 346:1133–4.

63. Ridker PM, Miletich JP, Hennekens CH, Buring JE. Ethnic distribution of factor V Leiden in 4047 men and women: implications for venous thromboembolism screening. JAMA 1997; 277:1305–7.

64. Rosendaal FR, Koster T, Vandenbroucke JP, Reitsma PH. High risk of thrombosis in patients homozygous for factor V Leiden (activated protein C resistance). Blood 1995; 85:1504–8.

65. Emmerich J, Rosendaal FR, Cattaneo M, et al. Combined effect of factor V Leiden and prothrombin 20210A on the risk of venous thromboembolism—pooled analysis of 8 case-controlled studies including 2310 cases and 3204 controls. Thromb Haemost 2001; 86:809–16. [Erratum, Thromb Haemost 2001; 86:1598.]

66. Vandenbroucke JP, Koster T, Briet E, Reitsma PH, Bertina RM, Rosendaal FR. Increased risk of venous thrombosis in oral-contraceptive users who are carriers of factor V Leiden mutation. Lancet 1994; 344:1453–7.

67. Vandenbroucke JP, van der Meer FJM, Helmerhorst FM, Rosendaal FR. Factor V Leiden: should we screen oral contraceptive users and pregnant women? BMJ 1996; 313:1127–30.

68. Grody WW, Griffin JH, Taylor AK, Korf BR, Heit JA. American College of Medical Genetics consensus statement on factor V Leiden mutation testing. Genet Med 2001; 3:139–48.

69. Committee on Genetics. Molecular genetic testing in pediatric practice: a subject review. Pediatrics 2000; 106:1494–7.

70. Nelson RM, Botkjin JR, Kodish ED, et al. Ethical issues with genetic testing in pediatrics. Pediatrics 2001; 107:1451–5.

71. The American Society of Human Genetics. Statement on informed consent for genetic research. Am J Hum Genet 1996; 59:471–4.

72. McCabe ERB. Clinical genetics: compassion, access, science, and advocacy. Genet Med 2001; 3:426–9.

73. Physicians' desk reference. 56th ed. Montvale, N.J.: Medical Economics, 2002.

5

Inheritance and Drug Response

RICHARD WEINSHILBOUM, M.D.

The promise of pharmacogenetics, the study of the role of inheritance in the individual variation in drug response, lies in its potential to identify the right drug and dose for each patient. Even though individual differences in drug response can result from the effects of age, sex, disease, or drug interactions, genetic factors also influence both the efficacy of a drug and the likelihood of an adverse reaction.[1–3] This chapter briefly reviews concepts that underlie the emerging fields of pharmacogenetics and pharmacogenomics, with an emphasis on the pharmacogenetics of drug metabolism. Although only a few examples will be provided to illustrate concepts and to demonstrate the potential contribution of pharmacogenetics to medical practice, it is now clear that virtually every pathway of drug metabolism will eventually be found to have genetic variation. The chapter by Evans and McLeod[4] expands on many of the themes introduced here.

Once a drug is administered, it is absorbed and distributed to its site of action, where it interacts with targets (such as receptors and enzymes), undergoes metabolism, and is then excreted.[5,6] Each of these processes could potentially involve clinically significant genetic variation. However, pharmacogenetics originated as a result of the observation that there are clinically important inherited variations in drug metabolism. Therefore, this chapter—and the examples highlighted—focuses on the pharmacogenetics of drug metabolism. However, similar principles apply to clinically significant inherited variation in the transport and distribution of drugs and their interaction with their therapeutic targets. The underlying message is that inherited variations in drug effect are common and that some tests that incorporate pharmacogenetics into clinical practice are now available, with many more to follow.

The concept of pharmacogenetics originated from the clinical observation that there were patients with very high or very low plasma or urinary drug concentrations, followed by the realization that the biochemical traits leading to this variation were inherited. Only later

Originally published February 6, 2003

were the drug-metabolizing enzymes identified, and this discovery was followed by the identification of the genes that encoded the proteins and the DNA-sequence variation within the genes that was associated with the inherited trait. Most of the pharmacogenetic traits that were first identified were monogenic—that is, they involved only a single gene—and most were due to genetic polymorphisms; in other words, the allele or alleles responsible for the variation were relatively common. Although drug effect is a complex phenotype that depends on many factors, early and often dramatic examples involving succinylcholine and isoniazid facilitated acceptance of the fact that inheritance can have an important influence on the effect of a drug. Today there is a systematic search to identify functionally significant variations in DNA sequences in genes that influence the effects of various drugs.[4]

PHARMACOGENETICS OF DRUG METABOLISM

Metabolism usually converts drugs to metabolites that are more water soluble and thus more easily excreted.[5] It can also convert prodrugs into therapeutically active compounds, and it may even result in the formation of toxic metabolites. Pharmacologists classify pathways of drug metabolism as either phase I reactions (i.e., oxidation, reduction, and hydrolysis) or phase II, conjugation reactions (e.g., acetylation, glucuronidation, sulfation, and methylation).[5] The names used to refer to these pathways for drug metabolism are purely historical, so phase II reactions can precede phase I reactions and often occur without prior oxidation, reduction, or hydrolysis. However, both types of reaction most often convert relatively lipid-soluble drugs into relatively more water-soluble metabolites (Fig. 5.1).

The finding, approximately 40 years ago, that an impairment in a phase I reaction—hydrolysis of the muscle relaxant succinylcholine by butyrylcholinesterase (pseudocholinesterase)—was inherited served as an early stimulus for the development of pharmacogenetics.[7] Approximately 1 in 3500 white subjects is homozygous for a gene encoding an atypical form of butyrylcholinesterase[8] and is relatively unable to hydrolyze succinylcholine, thus prolonging the drug-induced muscle paralysis and consequent apnea.[9] At almost the same time, it was observed that a common genetic variation in a phase II pathway of drug metabolism—N-acetylation—could result in striking differences in the half-life and plasma concentrations of drugs metabolized by N-acetyltransferase. Such drugs included the antituberculosis agent isoniazid,[10] the antihypertensive agent hydralazine,[11] and the antiarrhythmic drug procainamide,[12] and this variation had clinical consequences in all cases.[13] The bimodal distribution of plasma isoniazid concentrations in subjects with genetically determined fast or slow rates of acetylation in one of those early studies[10] strikingly illustrates the consequences of inherited variations in this pathway for drug metabolism (Fig. 5.2). These early examples of the potential influence of inheritance on the effect of a drug set the stage for subsequent studies of genetic variation in other pathways of drug biotransformation.

Pharmacogenetics of Phase I Drug Metabolism

The cytochrome P-450 enzymes, a superfamily of microsomal drug-metabolizing enzymes, are the most important of the enzymes that catalyze phase I drug metabolism.[5] One

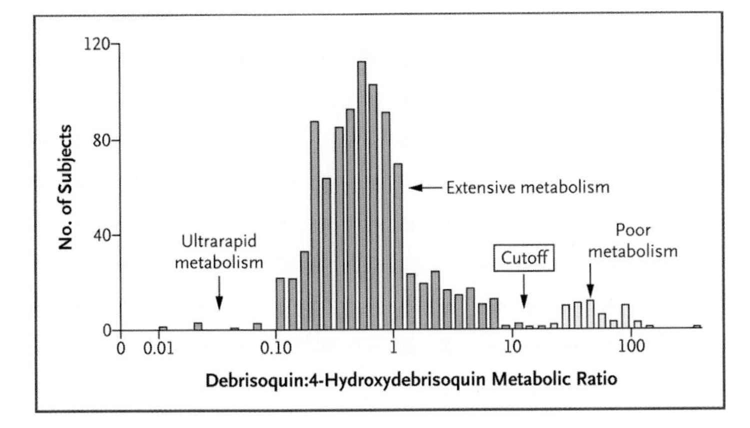

Figure 5.3 Pharmacogenetics of CYP2D6. Urinary metabolic ratios of debrisoquin to its metabolite, 4-hydroxydebrisoquin, are shown for 1011 Swedish subjects. The Cutoff box indicates the cutoff point between subjects with poor metabolism as a result of decreased or absent CYP2D6 activity and subjects with extensive metabolism. Modified from Bertilsson et al.[17] with the permission of the publisher.

laboratory. Therefore, the application of molecular genetic techniques to pharmacogenetics not only has made it possible to determine underlying molecular mechanisms responsible for genetic polymorphisms, but also has created the possibility of high-through-put clinical tests that can be performed with DNA isolated from a blood sample, an approach that is being adapted for routine diagnostic use in clinical laboratories.

Application of molecular genetic techniques resulted in the cloning of a complementary DNA (cDNA) and the gene encoding CYP2D6.[19,20] Those advances, in turn, made it possible to characterize a series of genetic variants responsible for low levels of CYP2D6 activity or no activity, ranging from single-nucleotide polymorphisms that altered the amino acid sequence of the encoded protein to single-nucleotide polymorphisms that altered RNA splicing or even deletions of the *CYP2D6* gene.[21] More than 75 *CYP2D6* alleles have now been described (descriptions are available at http://www.imm.ki.se/cypalleles). In addition, some subjects with ultrarapid metabolism have been shown to have multiple copies of the *CYP2D6* gene.[18] Such subjects can have an inadequate therapeutic response to standard doses of the drugs metabolized by CYP2D6. Although the occurrence of multiple copies of the *CYP2D6* gene is relatively infrequent among northern Europeans, in East African populations, the allele frequency can be as high as 29 percent.[22] The effect of the number of copies of the *CYP2D6* gene—ranging from 0 to 13—on the pharmacokinetics of the antidepressant drug nortriptyline is shown in Figure 5.4.[23] There could hardly be a more striking illustration of how genetics influences the metabolism of a drug.

The *CYP2D6* polymorphism represents an excellent example of both the potential clinical implications of pharmacogenetics and the process by which pharmacogenetic research led from the phenotype to an understanding of molecular mechanisms at the level of the genotype. Similar approaches were subsequently applied to other cytochrome P-450 isoforms, including 2C9, which metabolizes warfarin, losartan, and phenytoin; 2C19, which metabolizes omeprazole; and 3A5, which metabolizes a very large number of drugs.[24–26] We now know that many other phase I drug-metabolizing enzymes display genetic variation that can influence a person's response to a drug. Table 5.1 lists selected examples of clinically relevant pharmacogenetic variations involving phase I drug-metabolizing enzymes. In many cases, we also understand the molecular basis of inherited variation in the

Table 5.1 Pharmacogenetics of Phase I Drug Metabolism*

Drug-Metabolizing Enzyme	Frequency of Variant Poor-Metabolism Phenotype	Representative Drugs Metabolized	Effect of Polymorphism
Cytochrome P-450 2D6 (CYP2D6)	6.8% in Sweden 1% in China[17]	Debrisoquin[15] Sparteine[16] Nortriptyline[23] Codeine[27,28]	Enhanced drug effect Enhanced drug effect Enhanced drug effect Decreased drug effect
Cytochrome P-450 2C9 (CYP2C9)	Approximately 3% in England[29] (those homozygous for the *2 and *3 alleles)	Warfarin[29,30] Phenytoin[31,32]	Enhanced drug effect[29–32]
Cytochrome P-450 2C19 (CYP2C19)	2.7% among white Americans[33] 3.3% in Sweden 14.6% in China[17] 18% in Japan[33]	Omeprazole[34,35]	Enhanced drug effect[36,37]
Dihydropyrimidine dehydrogenase	Approximately 1% of population is heterozygous[38]	Fluorouracil[39,40]	Enhanced drug effect[39,40]
Butyrylcholinesterase (pseudocholinesterase)	Approximately 1 in 3500 Europeans[41]	Succinylcholine[9,41]	Enhanced drug effect[9,41]

*Examples of genetically polymorphic phase I enzymes are listed that catalyze drug metabolism, including selected examples of drugs that have clinically relevant variations in their effects.

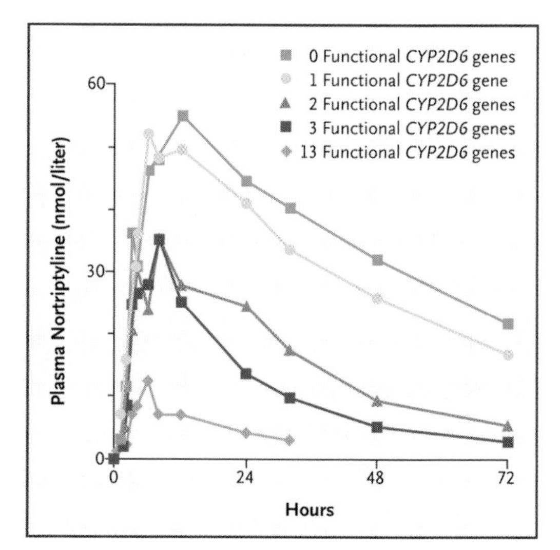

Figure 5.4 Pharmacogenetics of Nortriptyline. Mean plasma concentrations of nortriptyline after a single 25-mg oral dose are shown in subjects with 0, 1, 2, 3, or 13 functional *CYP2D6* genes. Modified from Dalén et al.[23] with the permission of the publisher.

drug-metabolizing enzymes. For example, in the atypical butyrylcholinesterase variant responsible for striking decreases in the ability to catalyze the hydrolysis of succinylcholine, guanine is substituted for adenine at position 209 in the open reading frame of the gene, resulting in a change from aspartic acid to glycine at position 70 in the encoded protein.[42] A series of other variant alleles for butyrylcholinesterase that result in decreased enzyme activity have also been described.[41]

Another example of the pharmacogenetics of phase I drug metabolism involves metabolism of the antineoplastic agent fluorouracil. In the mid-1980s, fatal central nervous system toxicity developed in several patients after treatment with standard doses of fluorouracil.[39,40] The patients were shown to have an inherited deficiency of dihydropyrimidine dehydrogenase, an enzyme that metabolizes fluorouracil and endogenous pyrimidines. Subsequently, several variant alleles for the gene encoding dihydropyrimidine dehydrogenase were described that placed patients at risk for toxic effects when they were exposed to standard doses of fluorouracil.[43] The pharmacogenetics of dihydropyrimidine dehydrogenase and its effect on the metabolism of fluorouracil, as well as the pharmacogenetics of thiopurine drugs discussed below, serve to illustrate another general principle: pharmacogenetic variation in the response to drugs has been recognized most often for drugs with narrow therapeutic indexes—drugs for which differences between the toxic and therapeutic doses are relatively small. However, the same general principles would be expected to apply to all therapeutic agents, and the same research strategies that were used to identify common, clinically significant genetic variations in phase I pathways of drug metabolism have also been applied—with equal success—to reactions involving phase II drug metabolism.

Pharmacogenetics of Phase II Drug Metabolism

The N-acetylation of isoniazid (Fig. 5.2) was an early example of inherited variation in phase II drug metabolism. Molecular cloning studies subsequently demonstrated that there are two N-acetyltransferase (*NAT*) genes in humans, *NAT1* and *NAT2*.[44] The common genetic polymorphism responsible for the pharmacogenetic variation in isoniazid metabolism illustrated in Figure 5.2 involved the *NAT2* gene. That polymorphism, like those in the genes for many other drug-metabolizing enzymes, shows striking ethnic variation.[45] As a result, most East Asian subjects are rapid acetylators of isoniazid and other drugs metabolized by N-acetyltransferase 2.[46] Although the *NAT2* genetic polymorphism was one of the earliest examples discovered of a pharmacogenetic variant in a phase II drug-metabolizing enzyme, it was a common genetic polymorphism involving another conjugating (i.e., phase II) enzyme that became one of the earliest clinically accepted pharmacogenetic tests.

The thiopurine drugs mercaptopurine and azathioprine—a prodrug that is converted to mercaptopurine in vivo—are purine antimetabolites used clinically as immunosuppressants and to treat neoplasias, such as acute lymphoblastic leukemia of childhood.[47] Thiopurines are metabolized in part by S-methylation catalyzed by the enzyme thiopurine S-methyltransferase (TPMT).[48,49] Approximately 20 years ago it was reported that white populations can be separated into three groups on the basis of the level of TPMT activity in their red cells and other tissues and that the level of activity was inherited in an autosomal codominant fashion (Fig. 5.5A).[50,51] Subsequently, it was shown that when persons who were homozygous for low levels of TPMT activity or for no activity (*TPMT^L/TPMT^L*) (Fig. 5.5A) received standard doses of thiopurines, they had greatly elevated concentrations of active metabolites, 6-thioguanine nucleotides, as well as a greatly increased risk of life-threatening, drug-induced myelosuppression.[52] As a result, the phenotypic test for the level of TPMT activity in red cells and, subsequently, DNA-based tests were among the first pharmacogenetic tests to be used in clinical practice. They are an example of the individualization of therapy on the basis of pharmacogenetic data. Patients with inherited low levels of TPMT activity can be treated with thiopurine drugs but only at greatly reduced doses, if drug-induced toxicity is to be avoided.[51] There is also evidence that in patients with very high levels of activity, the efficacy of thiopurine drugs is decreased,[53] presumably because the drugs are rapidly metabolized.

It is interesting to contrast the test used to determine the TPMT phenotype with that used originally to classify subjects as having either poor or extensive metabolism of CYP2D6. In the case of TPMT, a blood sample could be obtained and the enzymatic activity measured directly, whereas for CYP2D6 a probe drug had to be administered and a urine sample collected (Fig. 5.3). The fact that TPMT is expressed in an easily accessible cell—the red cell—facilitated the introduction of this pharmacogenetic test into clinical use. The availability of DNA-based tests means that the clinical application of pharmacogenetics could be greatly accelerated for a large number of genes that encode proteins important in drug response.

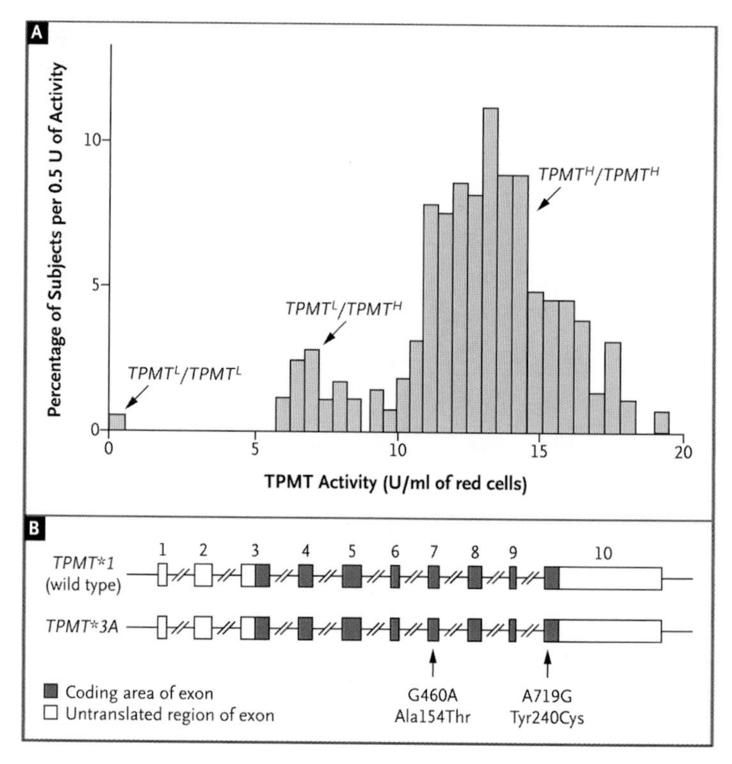

Figure 5.5 Pharmacogenetics of Thiopurine S-Methyltransferase (TPMT) (Panel A) and the *TPMT* Gene (Panel B). Panel A shows the level of TPMT activity in red cells among 298 randomly selected white adult blood donors. Presumed genotypes for the *TPMT* genetic polymorphism are also shown. *TPMT^L* and *TPMT^H* are alleles that result clinically in low levels and high levels of activity, respectively. These allele designations were used before the molecular basis for the polymorphism was understood. (Modified from Weinshilboum and Sladek[50] with the permission of the publisher.) Panel B shows the human *TPMT* gene. *TPMT*1* is the most common allele, and *TPMT*3A* is the most common variant allele among white subjects. The *TPMT*3A* allele is primarily responsible for the trimodal frequency distribution shown in Panel A. The two single-nucleotide polymorphisms in *TPMT*3A*, which are in strong linkage disequilibrium, as well as the resultant changes in encoded amino acids, are indicated.

The *TPMT* gene has been cloned,[54] and the most common variant allele responsible for low levels of activity among white populations encodes a protein with two alterations in the amino acid sequence as a result of single-nucleotide polymorphisms (Fig. 5.5B).[54,55] These sequence changes result in a striking reduction in the quantity of *TPMT*,[54] probably because the variant protein is degraded rapidly.[56] A series of less frequent *TPMT* variant alleles have also been described.[51] Finally, there are large differences in the types and frequencies of *TPMT* alleles among ethnic groups. For example, *TPMT**3A, the most common allele responsible for very low levels of enzyme activity in whites, with a frequency of approximately 4 percent (Fig. 5B), has not been observed in China, Korea, or Japan.[57] Other examples of pharmacogenetic variation involving phase II pathways of drug metabolism are listed in Table 5.2.

Recurring themes in pharmacogenetics include the presence of a few relatively common variant alleles of genes encoding proteins important in drug response, a larger number of much less frequent variant alleles, and striking differences in the types and frequencies of alleles among different populations and ethnic groups. These generalizations, based primarily on studies of drug metabolism, are now being extended to include common genetic variations in other proteins that might alter the effects of drug transporters or targets.

Most drugs are metabolized by several different enzymes, can be transported by other types of proteins, and ultimately interact with one or more targets. If several steps in this type of pathway were to display genetic variation—that is, if the effects were polygenic—clear multimodal frequency distributions like those shown in Figures 5.2 and 5.3 would quickly be replaced by multiple overlapping distributions. Therefore, even if inheritance

Table 5.2 Pharmacogenetics of Phase II Drug Metabolism*

Drug-Metabolizing Enzyme	Frequency of Variant Poor-Metabolism Phenotype	Representative Drugs Metabolized	Effect of Polymorphism
N-Acetyltransferase 2	52% among white Americans[10] 17% of Japanese[58]	Isoniazid[10] Hydralazine[11] Procainamide[12]	Enhanced drug effect[13]
Uridine diphosphate–glucuronosyltransferase 1A1 (TATA-box polymorphism)	10.9% among whites[59] 4% of Chinese[60] 1% of Japanese[60]	Irinotecan[61] Bilirubin[62]	Enhanced drug effect[63] Gilbert's syndrome[62]
Thiopurine *S*-methyltransferase	Approximately 1 in 300 whites[50,57] Approximately 1 in 2500 Asians[57]	Mercaptopurine[51] Azathioprine	Enhanced drug effect (toxicity)[51–53]
Catechol *O*-methyltransferase	Approximately 25% of whites[51,64]	Levodopa[51,65]	Enhanced drug effect[51,65]

*Examples of genetically polymorphic phase II (conjugating) enzymes are listed that catalyze drug metabolism, including selected examples of drugs that have clinically relevant variations in their effects.

influenced the effect of a drug, the relatively simple, one-to-one relation observed for CYP2D6 and TPMT would not be obvious. Perhaps partly as a result of that fact, there are still relatively few examples of clinical tests based on pharmacogenetics. An additional factor responsible for the relatively slow translation of pharmacogenetic approaches into clinical practice—as mentioned previously—has been the frequent requirement for the administration of a probe drug in the diagnosis of many pharmacogenetic traits. DNA-based assays make it possible to obtain pharmacogenetic information on a large number of genes encoding relevant proteins. The range of functionally significant variations in DNA sequences in genes that influence the response to drugs is wide and includes single-nucleotide polymorphisms, small insertions and deletions, variable-number tandem repeats, gene deletions, and gene duplications. And although DNA-based tests can be used to detect sequence variations, the results of such tests will not necessarily account for all possible phenotypic variations. What they can potentially do quickly is make available large quantities of data on many genes that might contribute to variations in drug response.

FROM PHARMACOGENETICS TO PHARMACOGENOMICS

The convergence of pharmacogenetics and rapid advances in human genomics has resulted in pharmacogenomics, a term used here to mean the influence of DNA-sequence variation on the effect of a drug. With the completion of the Human Genome Project[66,67] and the ongoing annotation of its data, the time is rapidly approaching when the sequences of virtually all genes encoding enzymes that catalyze phase I and II drug metabolism will be known. The same will be true for genes that encode drug transporters, drug receptors, and other drug targets. As a result, the traditional phenotype-to-genotype pharmacogenetic-research paradigm described at the beginning of this article is reversing direction to create a complementary genotype-to-phenotype flow of information.

CONCLUSIONS

The convergence of advances in pharmacogenetics and human genomics means that physicians can now individualize therapy in the case of a few drugs. As our knowledge of genetic variations in proteins involved in the uptake, distribution, metabolism, and action of various drugs improves, our ability to test for that variation and, as a result, to select the best drug at the optimal dose for each patient should also increase.

Acknowledgments

Supported in part by grants (RO1 GM28157, RO1 GM35720, and UO1 GM61388) from the National Institutes of Health.

I am indebted to Ms. Luanne Wussow for her assistance with the preparation of the manuscript.

When originally published, Dr. Weinshilboum reported providing consulting services to Abbott Laboratories, Bristol-Myers Squibb, Eli Lilly, and Johnson and Johnson; all fees for these services were paid to the Mayo Foundation.

References

1. Kalow W. Pharmacogenetics: heredity and the response to drugs. Philadelphia: W.B. Saunders, 1962.

2. Price Evans DA. Genetic factors in drug therapy: clinical and molecular pharmacogenetics. Cambridge, England: Cambridge University Press, 1993.

3. Weber WW. Pharmacogenetics. New York: Oxford University Press, 1997.

4. Evans WE, McLeod HL. Pharmacogenomics—drug disposition, drug targets, and side effects. N Engl J Med 2003; 348:538–49.

5. Wilkinson GR. Pharmacokinetics: the dynamics of drug absorption, distribution, and elimination. In: Hardman JG, Limbird LE, Gilman AG, eds. Goodman & Gilman's the pharmacological basis of therapeutics. 10th ed. New York: McGraw-Hill, 2001:3–29.

6. Ross EM, Kenakin TP. Pharmacodynamics: mechanisms of drug action and the relationship between drug concentration and effect. In: Hardman JG, Limbird LE, Gilman AG, eds. Goodman & Gilman's the pharmacological basis of therapeutics. 10th ed. New York: McGraw-Hill, 2001:31–43.

7. Kalow W. The Pennsylvania State University College of Medicine 1990 Bernard B. Brodie Lecture: pharmacogenetics: past and future. Life Sci 1990; 47:1385–97.

8. Kalow W, Gunn DR. Some statistical data on atypical cholinesterase of human serum. Ann Hum Genet 1959; 23:239–50.

9. Idem. The relation between dose of succinylcholine and duration of apnea in man. J Pharmacol Exp Ther 1957; 120:203–14.

10. Price Evans DA, Manley KA, McKusick VA. Genetic control of isoniazid metabolism in man. BMJ 1960; 2:485–91.

11. Timbrell JA, Harland SJ, Facchini V. Polymorphic acetylation of hydralazine. Clin Pharmacol Ther 1980; 28:350–5.

12. Reidenberg MM, Drayer DE, Levy M, Warner H. Polymorphic acetylation of procainamide in man. Clin Pharmacol Ther 1975; 17:722–30.

13. Drayer DE, Reidenberg MM. Clinical consequences of polymorphic acetylation of basic drugs. Clin Pharmacol Ther 1977; 22:251–8.

14. Kroemer HK, Eichelbaum M. "It's the genes, stupid": molecular bases and clinical consequences of genetic cytochrome P450 2D6 polymorphism. Life Sci 1995; 56:2285–98.

15. Mahgoub A, Idle JR, Dring LG, Lancaster R, Smith RL. Polymorphic hydroxylation of debrisoquine in man. Lancet 1977; 2:584–6.

16. Eichelbaum M, Spannbrucker N, Steincke B, Dengler HJ. Defective N-oxidation of sparteine in man: a new pharmacogenetic defect. Eur J Clin Pharmacol 1979; 16:183–7.

17. Bertilsson L, Lou YQ, Du YL, et al. Pronounced differences between native Chinese and Swedish populations in the polymorphic hydroxylations of debrisoquin and S-mephenytoin. Clin Pharmacol Ther 1992; 51:388–97. [Erratum, Clin Pharmacol Ther 1994; 55:648.]

18. Johansson I, Lundqvist E, Bertilsson L, Dahl ML, Sjoqvist F, Ingelman-Sundberg M. Inherited amplification of an active gene in the cytochrome P450 CYP2D locus as a cause of ultrarapid metabolism of debrisoquine. Proc Natl Acad Sci U S A 1993; 90: 11825–9.

19. Gonzalez FJ, Vilbois F, Hardwick JP, et al. Human debrisoquine 4-hydroxylase (P450IID1): cDNA and deduced amino acid sequence and assignment of the CYP2D locus of chromosome 22. Genomics 1988; 2:174–9.

20. Kimura S, Umeno M, Skoda R, Meyer UA, Gonzalez FJ. The human debrisoquine 4-hydroxylase (CYP2D) locus: sequence and identification of the polymorphic CYP2D6 gene, a related gene and a pseudogene. Am J Hum Genet 1989; 45:889–904.

21. Ingelman-Sundberg M, Evans WE. Unravelling the functional genomics of the human CYP2D6 gene locus. Pharmacogenetics 2001; 11:553–4.

22. Aklillu E, Persson I, Bertilsson L, Johansson I, Rodrigues F, Ingelman-Sundberg M. Frequent distribution of ultrarapid metabolizers of debrisoquine in an Ethiopian population carrying duplicated and multiduplicated functional CYP2D6 alleles. J Pharmacol Exp Ther 1996; 278:441–6.

23. Dalén P, Dahl ML, Ruiz MLB, Nordin J, Bertilsson L. 10-Hydroxylation of nortriptyline in white persons with 0, 1, 2, 3, and 13 functional CYP2D6 genes. Clin Pharmacol Ther 1998; 63:444–52.

24. Stubbins MJ, Harris LW, Smith G, Tarbit MH, Wolf CR. Genetic analysis of the human cytochrome P450 CYP2C9 locus. Pharmacogenetics 1996; 6:429–39.

25. de Morais SMF, Wilkinson GR, Blaisdell J, Nakamura K, Meyer UA, Goldstein JA. The major genetic defect responsible for the polymorphism of S-mephenytoin metabolism in humans. J Biol Chem 1994; 269:15419–22.

26. Kuehl P, Zhang J, Lin Y, et al. Sequence diversity in CYP3A promoters and characterization of the genetic basis of polymorphic CYP3A5 expression. Nat Genet 2001; 27:383–91.

27. Mortimer O, Persson K, Ladona MG, et al. Polymorphic formation of morphine from codeine in poor and extensive metabolizers of dextromethorphan: relationship to the presence of immunoidentified

cytochrome P-450IID1. Clin Pharmacol Ther 1990; 47:27–35.

28. Sindrup SH, Brøsen K. The pharmacogenetics of codeine hypoalgesia. Pharmacogenetics 1995; 5:335–46.

29. Aithal GP, Day CP, Kesteven PJL, Daly AK. Association of polymorphisms in the cytochrome P450 CYP2C9 with warfarin dose requirement and risk of bleeding complications. Lancet 1999; 353:717–9.

30. Loebstein R, Yonath H, Peleg D, et al. Interindividual variability in sensitivity to warfarin—nature or nurture? Clin Pharmacol Ther 2001; 70:159–64.

31. van der Weide J, Steijns LSW, van Weelden MJM, de Haan K. The effect of genetic polymorphism of cytochrome P450 CYP2C9 on phenytoin dose requirement. Pharmacogenetics 2001; 11:287–91.

32. Brandolese R, Scordo MG, Spina E, Gusella M, Padrini R. Severe phenytoin intoxication in a subject homozygous for CYP2C9*3. Clin Pharmacol Ther 2001; 70:391–4.

33. Nakamura K, Goto F, Ray WA, et al. Interethnic differences in genetic polymorphism of debrisoquin and mephenytoin hydroxylation between Japanese and Caucasian populations. Clin Pharmacol Ther 1985; 38:402–8.

34. Balian J, Sukhova N, Harris JW, et al. The hydroxylation of omeprazole correlates with S-mephenytoin metabolism: a population study. Clin Pharmacol Ther 1995; 57:662–9.

35. Tybring G, Böttiger Y, Widén J, Bertilsson L. Enantioselective hydroxylation of omeprazole catalyzed by CYP2C19 in Swedish white subjects. Clin Pharmacol Ther 1997; 62:129–37.

36. Furuta T, Ohashi K, Kamata T, et al. Effect of genetic differences in omeprazole metabolism on cure rate for *Helicobacter pylori* infection and peptic ulcer. Ann Intern Med 1998; 129:1027–30.

37. Tanigawara Y, Aoyama N, Kita T, et al. *CYP2C19* genotype-related efficacy of omeprazole for the treatment of infection caused by *Helicobacter pylori*. Clin Pharmacol Ther 1999; 66:528–34.

38. Raida M, Schwabe W, Hausler P, et al. Prevalence of a common point mutation in the dihydropyrimidine dehydrogenase (DPD) gene within the 5'-splice donor site of intron 14 in patients with severe 5-fluorouracil (5-FU)-related toxicity compared with controls. Clin Cancer Res 2001; 7:2832–9.

39. Tuchman M, Stoeckeler JS, Kiang DT, O'Dea RF, Ramnaraine ML, Mirkin BL. Familial pyrimidinemia and pyrimidinurea associated with severe fluorouracil toxicity. N Engl J Med 1985; 313:245–9.

40. Diasio RB, Beavers TL, Carpenter JT. Familial deficiency of dihydropyrimidine dehydrogenase: biochemical basis for familial pyrimidinemia and severe 5-fluorouracil-induced toxicity. J Clin Invest 1988; 81:47–51.

41. Lockridge O. Genetic variants of human butyrylcholinesterase influence the metabolism of the muscle relaxant succinylcholine. In: Kalow W, ed. Pharmacogenetics of drug metabolism. International encyclopedia of pharmacology and therapeutics. New York: Pergamon Press, 1992:15–50.

42. McGuire MC, Nogueira CP, Bartels CF, et al. Identification of the structural mutation responsible for the dibucaine-resistant (atypical) variant form of human serum cholinesterase. Proc Natl Acad Sci U S A 1989; 86:953–7.

43. McLeod HL, Collie-Duguid ESR, Vreken P, et al. Nomenclature for human DPYD alleles. Pharmacogenetics 1998; 8:455–9.

44. Blum M, Grant DM, McBride W, Heim M, Meyer UA. Human arylamine N-acetyltransferase genes: isolation, chromosomal localization, and functional expression. DNA Cell Biol 1990; 9:193–203.

45. Lin HJ, Han C-Y, Lin BK, Hardy S. Slow acetylator mutations in the human polymorphic N-acetyltransferase gene in 786 Asians, blacks, Hispanics, and whites: application to metabolic epidemiology. Am J Hum Genet 1993; 52:827–34.

46. Price Evans DA. N-acetyltransferase in pharmacogenetics of drug metabolism. In: Kalow W, ed. Pharmacogenetics of drug metabolism. International encyclopedia of pharmacology and therapeutics. New York: Pergamon Press, 1992:43:95–178.

47. Lennard L. The clinical pharmacology of 6-mercaptopurine. Eur J Clin Pharmacol 1992; 43:329–39.

48. Remy CN. Metabolism of thiopyrimidines and thiopurines: S-methylation with S-adenosylmethionine transmethylase and catabolism in mammalian tissues. J Biol Chem 1963; 238:1078–84.

49. Woodson LC, Weinshilboum RM. Human kidney thiopurine methyltransferase: purification and biochemical properties. Biochem Pharmacol 1983; 32:819–26.

50. Weinshilboum RM, Sladek SL. Mercaptopurine pharmacogenetics: monogenic inheritance of erythrocyte thiopurine methyltransferase activity. Am J Hum Genet 1980; 32:651–62.

51. Weinshilboum RM, Otterness DM, Szumlanski CL. Methylation pharmacogenetics: catechol O-methyltransferase, thiopurine methyltransferase, and histamine N-methyltransferase. Annu Rev Pharmacol Toxicol 1999; 39:19–52.

52. Lennard L, Van Loon JA, Weinshilboum RM. Pharmacogenetics of acute azathioprine toxicity:

relationship to thiopurine methyltransferase genetic polymorphism. Clin Pharmacol Ther 1989; 46:149–54.

53. Lennard L, Lilleyman JS, Van Loon J, Weinshilboum RM. Genetic variation in response to 6-mercaptopurine for childhood acute lymphoblastic leukaemia. Lancet 1990; 336:225–9.

54. Szumlanski C, Otterness D, Her C, et al. Thiopurine methyltransferase pharmacogenetics: human gene cloning and characterization of a common polymorphism. DNA Cell Biol 1996; 15:17–30.

55. Tai H-L, Krynetski EY, Yates CR, et al. Thiopurine S-methyltransferase deficiency: two nucleotide transitions define the most prevalent mutant allele associated with loss of catalytic activity in Caucasians. Am J Hum Genet 1996; 58:694–702.

56. Tai H-L, Krynetski EY, Schuetz EG, Yanishevski Y, Evans WE. Enhanced proteolysis of thiopurine S-methyltransferase (TPMT) encoded by mutant alleles in humans *(TPMT**3A, *TPMT**2): mechanisms for the genetic polymorphism of TPMT activity. Proc Natl Acad Sci U S A 1997; 94:6444–9.

57. Collie-Duguid ESR, Pritchard SC, Powrie RH, et al. The frequency and distribution of thiopurine methyltransferase alleles in Caucasian and Asian populations. Pharmacogenetics 1999; 9:37–42.

58. Dufour AP, Knight RA, Harris HW. Genetics of isoniazid metabolism in Caucasian, Negro, and Japanese populations. Science 1964; 145:391.

59. Lampe JW, Bigler J, Horner NK, Potter JD. UDP-glucuronosyltransferase (UGT1A1*28 and UGT1A6*2) polymorphisms in Caucasians and Asians: relationships to serum bilirubin concentrations. Pharmacogenetics 1999; 9:341–9.

60. Hall D, Ybazeta G, Destro-Bisol G, Petzl-Erler ML, Di Rienzo A. Variability at the uridine diphosphate glucuronosyltransferase 1A1 promoter in human populations and primates. Pharmacogenetics 1999; 9:591–9.

61. Iyer L, Hall D, Das S, et al. Phenotype-genotype correlation of in vitro SN-38 (active metabolite of irinotecan) and bilirubin glucuronidation in human liver tissue with UGT1A1 promoter polymorphism. Clin Pharmacol Ther 1999; 65:576–82.

62. Bosma PJ, Chowdhury JR, Bakker C, et al. The genetic basis of the reduced expression of bilirubin UDP-glucuronosyltransferase 1 in Gilbert's syndrome. N Engl J Med 1995; 333:1171–5.

63. Iyer L, Das S, Janisch L, et al. *UGT1A1**28 polymorphism as a determinant of irinotecan disposition and toxicity. Pharmacogenetics J 2002; 2:43–7.

64. Weinshilboum RM, Raymond FA. Inheritance of low erythrocyte catechol-O-methyltransferase activity in man. Am J Hum Genet 1977; 29:125–35.

65. Reilly DK, Rivera-Calimlim L, Van Dyke D. Catechol-O-methyltransferase activity: a determinant of levodopa response. Clin Pharmacol Ther 1980; 28:278–86.

66. Lander ES, Linton LM, Birren B, et al. Initial sequencing and analysis of the human genome. Nature 2001; 409:860–921. [Errata, Nature 2001; 411:720, 412:565.]

67. Venter JC, Adams MD, Myers EW, et al. The sequence of the human genome. Science 2001; 291:1304–51. [Erratum, Science 2001; 292:1838.]

6

Pharmacogenomics—Drug Disposition, Drug Targets, and Side Effects

WILLIAM E. EVANS, Pharm.D., and HOWARD L. McLEOD, Pharm.D.

It is well recognized that different patients respond in different ways to the same medication. These differences are often greater among members of a population than they are within the same person at different times (or between monozygotic twins).[1] The existence of large population differences with small intrapatient variability is consistent with inheritance as a determinant of drug response; it is estimated that genetics can account for 20 to 95 percent of variability in drug disposition and effects.[2] Although many nongenetic factors influence the effects of medications, including age, organ function, concomitant therapy, drug interactions, and the nature of the disease, there are now numerous examples of cases in which interindividual differences in drug response are due to sequence variants in genes encoding drug-metabolizing enzymes, drug transporters, or drug targets.[3–5] Unlike other factors influencing drug response, inherited determinants generally remain stable throughout a person's lifetime.

Clinical observations of inherited differences in drug effects were first documented in the 1950s,[6–9] giving rise to the field of pharmacogenetics, and later pharmacogenomics. Although the two terms are synonymous for all practical purposes, pharmacogenomics uses genome-wide approaches to elucidate the inherited basis of differences between persons in the response to drugs.

More than 1.4 million single-nucleotide polymorphisms were identified in the initial sequencing of the human genome,[10] with over 60,000 of them in the coding region of genes. Some of these single-nucleotide polymorphisms have already been associated with substantial changes in the metabolism or effects of medications, and some are now being used to predict clinical response.[3–5,11] Because most drug effects are determined

Originally published February 6, 2003

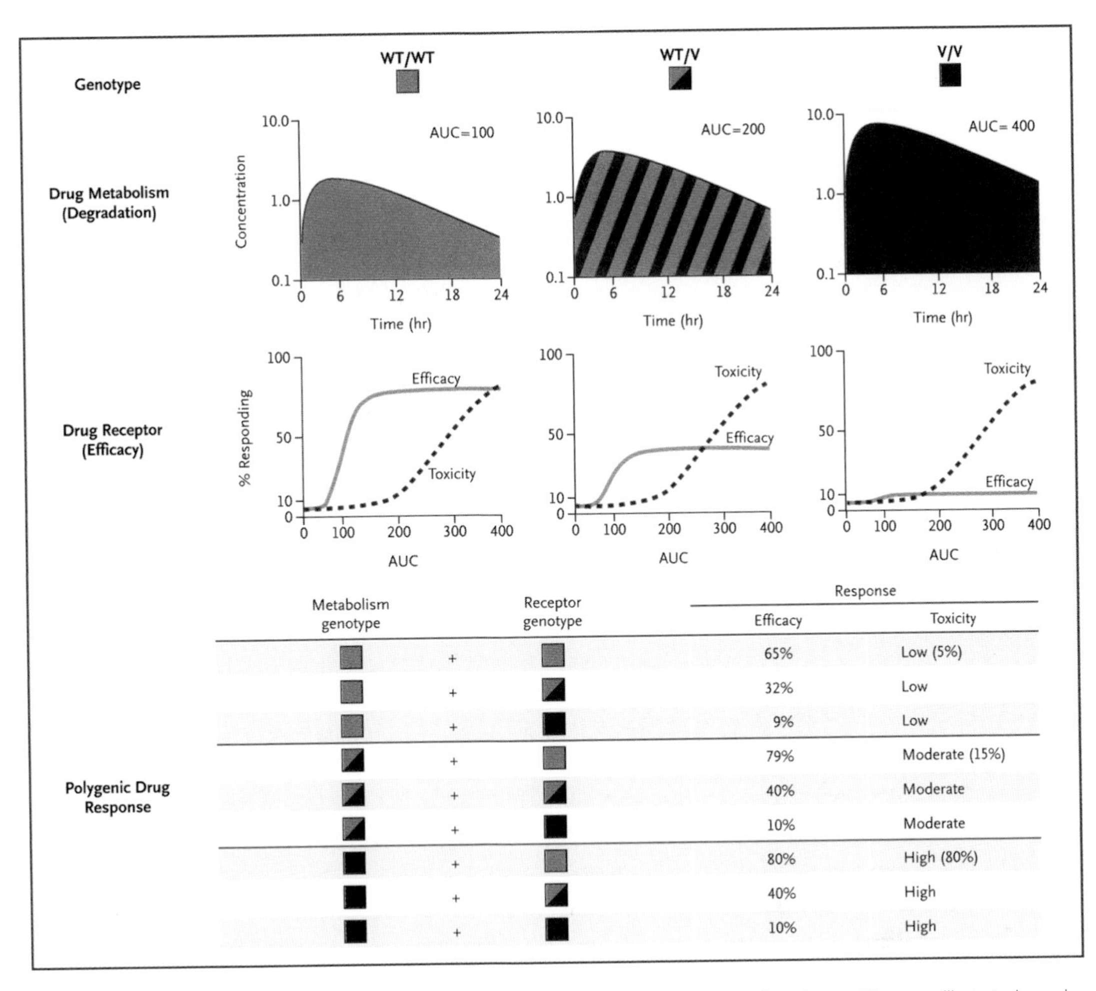

Figure 6.1 Polygenic Determinants of Drug Response. The potential effects of two genetic polymorphisms are illustrated, one involving a drug-metabolizing enzyme (top) and the second involving a drug receptor (middle), depicting differences in drug clearance (or the area under the plasma concentration–time curve [AUC]) and receptor sensitivity in patients who are homozygous for the wild-type allele (WT/WT), are heterozygous for one wild-type and one variant (V) allele (WT/V), or have two variant alleles (V/V) for the two polymorphisms. At the bottom are shown the nine potential combinations of drug-metabolism and drug-receptor genotypes and the corresponding drug-response phenotypes calculated from data at the top, yielding therapeutic indexes (efficacy:toxicity ratios) ranging from 13 (65 percent:5 percent) to 0.125 (10 percent:80 percent).

by the interplay of several gene products that influence the pharmacokinetics and pharmacodynamics of medications, including inherited differences in drug targets (e.g., receptors) and drug disposition (e.g., metabolizing enzymes and transporters), polygenic determinants of drug effects (Fig. 6.1) have become increasingly important in pharmacogenomics. In this chapter, we focus on the therapeutic consequences of inherited differences in drug disposition and drug targets. Chapter 7 focuses on the pharmacogenetics of drug metabolism. This review is not meant to be exhaustive; rather, clinically relevant examples are used to illustrate how pharmacogenomics can provide molecular diagnostic methods that improve drug therapy.

The field of pharmacogenetics began with a focus on drug metabolism,[12] but it has been extended to encompass the full spectrum of drug disposition, including a growing list of transporters that influence drug absorption, distribution, and excretion.[3–5,13]

Drug Metabolism

There are more than 30 families of drug-metabolizing enzymes in humans,[3,14] and essentially all have genetic variants, many of which translate into functional changes in the proteins encoded. These monogenic traits are discussed by Weinshilboum.[12] But there is an instructive example of a multigenic effect involving the CYP3A family of P-450 enzymes. About three quarters of whites and half of blacks have a genetic inability to express functional CYP3A5.[15] The lack of functional CYP3A5 may not be readily evident, because many medications metabolized by CYP3A5 are also metabolized by the universally expressed CYP3A4. For medications that are equally metabolized by both enzymes, the net rate of metabolism is the sum of that due to CYP3A4 and that due to CYP3A5; the existence of this dual pathway partially obscures the clinical effects of genetic polymorphism of CYP3A5 but contributes to the large range of total CYP3A activity in humans (Fig. 6.2). The CYP3A pathway of drug elimination is further confounded by the presence of single-nucleotide polymorphisms in the *CYP3A4* gene that alter the activity of this enzyme for some substrates but not for others.[16] The genetic basis of CYP3A5 deficiency is predominantly a single-nucleotide polymorphism

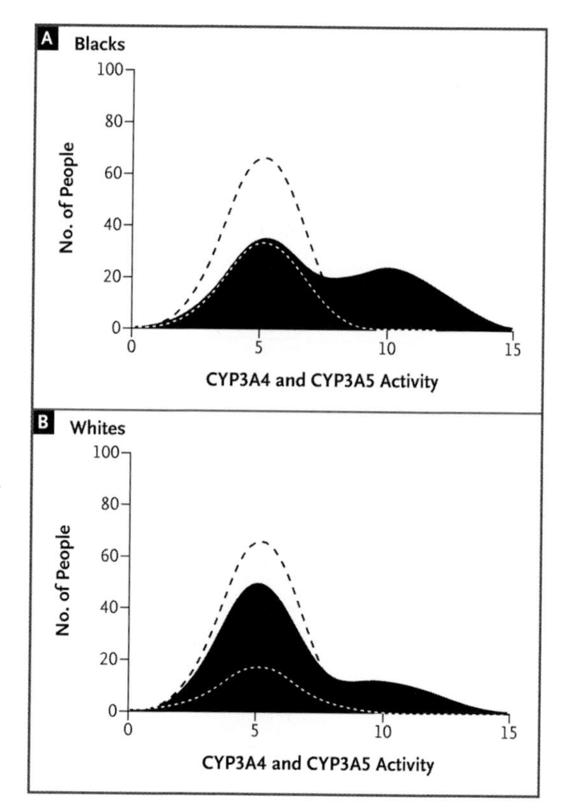

Figure 6.2 Simulated Activities of Cytochromes P-450 CYP3A4 and CYP3A5 in Blacks and Whites. The simulated activities of CYP3A4 (black dashed lines) and CYP3A5 (white dashed lines) are shown in blacks (Panel A) and whites (Panel B), assuming a normal distribution and a 10-fold range in activity (shown in arbitrary units) among those expressing functional forms of these enzymes, and further assuming that all patients express CYP3A4, but that only 25 percent of whites and 50 percent of blacks express functional CYP3A5 because of genetic polymorphism. The solid area reflects the combined activity of CYP3A4 and CYP3A5 in the two populations for medications that are metabolized equally by the two enzymes.

in intron 3 that creates a cryptic splice site causing 131 nucleotides of the intronic sequence to be inserted into the RNA, introducing a termination codon that prematurely truncates the CYP3A5 protein.[15] Although it is now possible to determine which patients express both functional enzymes (i.e., CYP3A4 and CYP3A5), the clinical importance of these variants for the many drugs metabolized by CYP3A remains unclear.

Drug Transporters

Transport proteins have an important role in regulating the absorption, distribution, and excretion of many medications. Members of the adenosine triphosphate (ATP)–binding cassette family of membrane transporters[17] are among the most extensively studied transporters involved in drug disposition and effects. A member of the ATP-binding cassette family, P-glycoprotein, is encoded by the human *ABCB1* gene (also called *MDR1*). A principal function of P-glycoprotein is the energy-dependent cellular efflux of substrates, including bilirubin, several anticancer drugs, cardiac glycosides, immunosuppressive agents, glucocorticoids, human immunodeficiency virus (HIV) type 1 protease inhibitors, and many other medications (Fig. 6.3).[17,21,22] The expression of P-glycoprotein in many normal tissues suggests that it has a role in the excretion of xenobiotics and metabolites into urine, bile, and the intestinal lumen.[23,24] At the blood–brain barrier, P-glycoprotein in the choroid plexus limits the accumulation of many drugs in the brain, including digoxin, ivermectin, vinblastine, dexamethasone, cyclosporine, domperidone, and loperamide.[23–25] A synonymous single-nucleotide polymorphism (i.e., a single-nucleotide polymorphism that does not alter the amino acid encoded) in exon 26 (3435C → T) has been associated with variable expression of P-glycoprotein in the duodenum; in patients homozygous for the T allele, duodenal expression of P-glycoprotein was less than half that in patients with the CC genotype.[19] CD56+ natural killer cells from subjects homozygous for 3435C demonstrated significantly lower ex vivo retention of the P-glycoprotein substrate rhodamine (i.e., higher P-glycoprotein function).[26] Digoxin, another P-glycoprotein substrate, has significantly higher bioavailability in subjects with the 3435TT genotype.[19,27] As is typical for many pharmacogenetic traits, there is considerable racial variation in the frequency of the 3435C → T single-nucleotide polymorphism.[28–30]

The 3435C → T single-nucleotide polymorphism is in linkage disequilibrium with a nonsynonymous single-nucleotide polymorphism (i.e., one causing an amino acid change) in exon 21 (2677G → T, leading to Ala893Ser) that alters P-glycoprotein function.[18] Because these two single-nucleotide polymorphisms travel together, it is unclear whether the 3435C → T polymorphism is of functional importance or is simply linked with the causative polymorphism in exon 21. The 2677G → T single-nucleotide polymorphism has been associated with enhanced P-glycoprotein function in vitro and lower plasma fexofenadine concentrations in humans,[18] effects opposite to those reported with digoxin.[27]

The associations between treatment outcome and genetic variants in *CYP3A4*, *CYP3A5*, *CYP2D6*, *CYP2C19*, the chemokine receptor gene *CCR5*, and *ABCB1* have been examined in HIV-infected patients receiving combination antiretroviral therapy with either a protease inhibitor or a nonnucleoside reverse-transcriptase inhibitor.[20] The *ABCB1* 3435C → T polymorphism was associated with significant differences in the

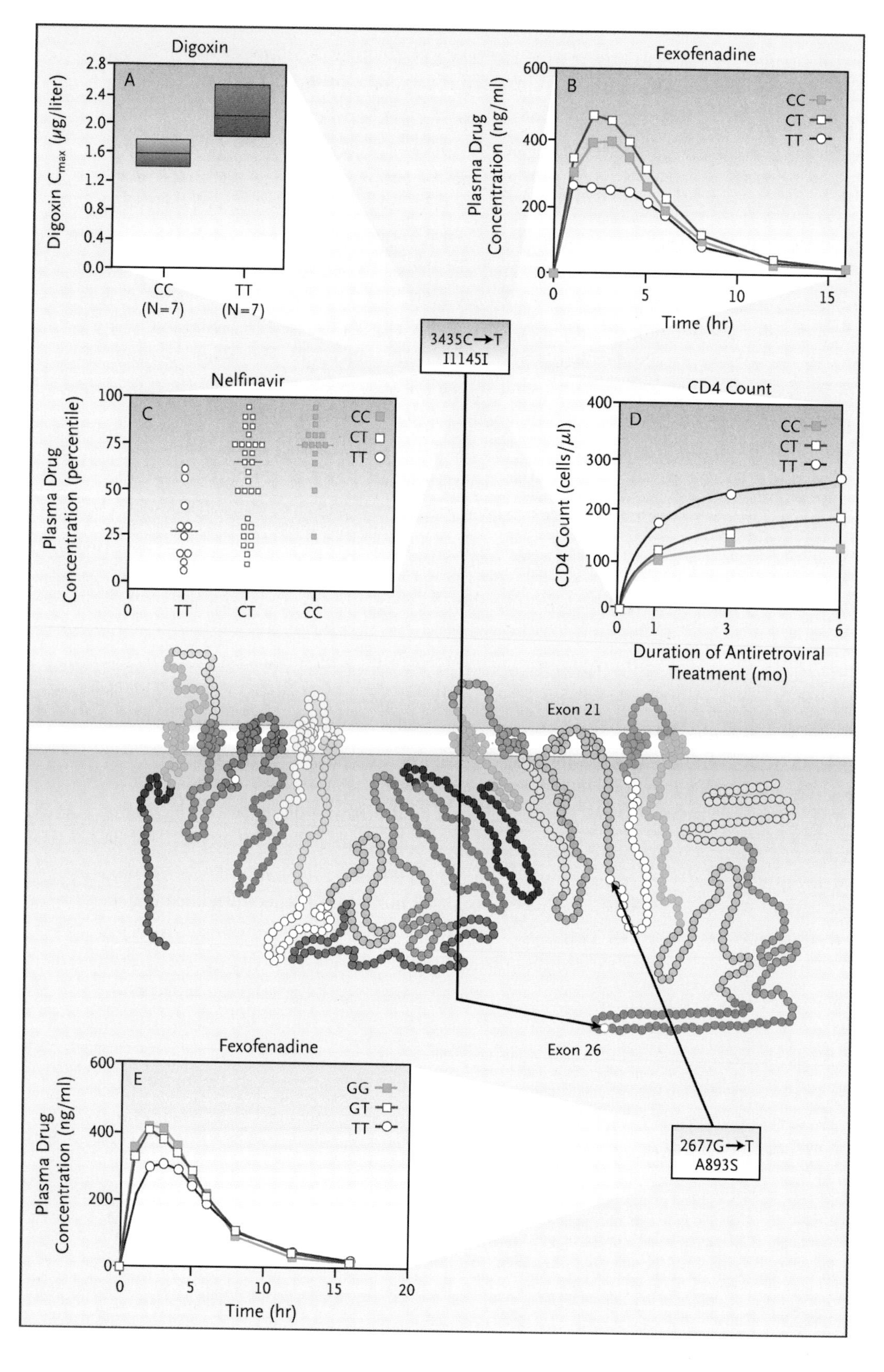

Digoxin
A
Digoxin C_{max} (μg/liter)
2.8
2.4
2.0
1.6
1.2
0.8
0.4
0.0
CC (N=7)
TT (N=7)
Fexofenadine
B
Plasma Drug Concentration (ng/ml)
600
400
200
0
CC
CT
TT
0
5
10
15
Time (hr)
3435C→T
I1145I
Nelfinavir
C
Plasma Drug Concentration (percentile)
100
75
50
25
0
TT
CT
CC
CD4 Count
D
CD4 Count (cells/μl)
400
300
200
100
0
1
3
6
Duration of Antiretroviral Treatment (mo)
Exon 21
Exon 26
Fexofenadine
E
Plasma Drug Concentration (ng/ml)
GG
GT
TT
20
Time (hr)
2677G→T
A893S

plasma pharmacokinetics of nelfinavir (Fig. 6.3) and efavirenz. Recovery of the CD4 cell count was significantly greater and more rapid in patients with the TT genotype than in patients with either the CT or the CC genotype (Fig. 6.3). Of many variables evaluated, only the *ABCB1* genotype and the base-line number of HIV RNA copies were significant predictors of CD4 recovery.[20] However, the *ABCB1* 2677G → T single-nucleotide polymorphism was not genotyped, so it remains unclear whether the 3435C → T polymorphism is causative or is simply linked with another polymorphism that is causative.

This example illustrates a common problem in association studies, namely, biologic plausibility. It is not obvious how greater efficacy (CD4 recovery) could be linked to a single-nucleotide polymorphism associated with lower plasma drug concentrations, unless there are specific effects of the *ABCB1* polymorphisms that cause decreased drug efflux from CD4 leukocytes. Overexpression of the gene for another ABC transporter (*ABCC4*, or *MRP4*) confers resistance to some nucleoside antiretroviral agents (e.g., zidovudine).[31] Despite the uncertainty about the mechanisms involved, the clinical value is that a host genetic marker can predict immune recovery after the initiation of antiretroviral treatment and, if validated, may offer a new strategy in tailoring HIV therapy.

GENETIC POLYMORPHISM OF DRUG TARGETS

Genetic variation in drug targets (e.g., receptors) can have a profound effect on drug efficacy, with over 25 examples already identified (Table 6.1).[3–5] Sequence variants with a direct effect on response occur in the gene for the β_2-adrenoreceptor, affecting the response to β_2-agonists[43,44]; arachidonate 5-lipoxygenase (ALOX5), affecting the response to ALOX5 inhibitors[42]; and angiotensin-converting enzyme (ACE), affecting the renoprotective actions of ACE inhibitors.[32] Genetic differences may also have indirect effects on drug response that are unrelated to drug metabolism or transport, such as methylation of the methylguanine methyltransferase (MGMT) gene promoter, which alters the response of gliomas to treatment with carmustine.[63] The mechanism of this effect is related to a decrease in the efficiency of repair of alkylated DNA in patients with methylated MGMT. It is critical to distinguish this target mechanism from genetic polymorphisms in drug-metabolizing enzymes that affect response by altering drug concentrations, such as the

Figure 6.3 (facing page). Functional Consequences of Genetic Polymorphisms in the Human P-Glycoprotein Transporter Gene *ABCB1* (or *MDR1*). The schematic diagram of the human P-glycoprotein was adapted from Kim et al.,[18] with each circle representing an amino acid and each color a different exon encoding the corresponding amino acids. Two single-nucleotide polymorphisms in the human *ABCB1* gene have been associated with altered drug disposition (Panels A, B, C, and E) or altered drug effects (Panel D). The synonymous single-nucleotide polymorphism (a single-nucleotide polymorphism that does not alter the amino acid encoded) in exon 26 (the 3435C → T single-nucleotide polymorphism) has been associated with higher oral bioavailability of digoxin in patients homozygous for the T nucleotide (Panel A [C_{max} denotes maximal concentration])[19] but lower plasma concentrations after oral doses of fexofenadine (Panel B)[18] and nelfinavir (Panel C).[20] This single-nucleotide polymorphism has also been linked to better CD4 cell recovery in HIV-infected patients who are treated with nelfinavir and other antiretroviral agents (Panel D).[20] The single-nucleotide polymorphism at nucleotide 2677 (G → T) has been associated with lower plasma fexofenadine concentrations in patients homozygous for the T nucleotide at position 2677 (Panel E).[18] The panels have been adapted from Kim et al.,[18] Hoffmeyer et al.,[19] and Fellay et al.[20]

Table 6.1 Genetic Polymorphisms in Drug Target Genes That Can Influence Drug Response*

Gene or Gene Product	Medication	Drug Effect Associated with Polymorphism
ACE	ACE inhibitors (e.g., enalapril)	Renoprotective effects, blood-pressure reduction, reduction in left ventricular mass, endothelial function[32-40]
	Fluvastatin	Lipid changes (e.g., reductions in low-density lipoprotein cholesterol and apolipoprotein B); progression or regression of coronary atherosclerosis[41]
Arachidonate 5-lipoxygenase	Leukotriene inhibitors	Improvement in FEV_1[42]
β_2-Adrenergic receptor	β_2-Agonists (e.g., albuterol)	Bronchodilatation, susceptibility to agonist-induced desensitization, cardiovascular effects[43–50]
Bradykinin B2 receptor	ACE inhibitors	ACE-inhibitor–induced cough[51]
Dopamine receptors (D2, D3, D4)	Antipsychotics (e.g., haloperidol, clozapine)	Antipsychotic response (D2, D3, D4), antipsychotic-induced tardive dyskinesia (D3), antipsychotic-induced acute akathisia (D3)[52–56]
Estrogen receptor-α	Conjugated estrogens Hormone-replacement therapy	Increase in bone mineral density[57] Increase in high-density lipoprotein cholesterol[58]
Glycoprotein IIIa subunit of glycoprotein IIb/IIIa	Aspirin or glycoprotein IIb/IIIa inhibitors	Antiplatelet effect[59]
Serotonin (5-hydroxytryptamine) transporter	Antidepressants (e.g., clomipramine, fluoxetine, paroxetine)	5-Hydroxytryptamine neurotransmission, antidepressant response[60–62]

*The examples shown are illustrative and not representative of all published studies, which exceed the scope of this review. ACE denotes angiotensin-converting enzyme, and FEV_1 forced expiratory volume in one second.

thiopurine methyltransferase polymorphism associated with the hematopoietic toxicity of mercaptopurine[64–66] and susceptibility to radiation-induced brain tumors.[67]

The β_2-adrenoreceptor (coded by the *ADRB2* gene) illustrates another link between genetic polymorphisms in drug targets and clinical responses. Genetic polymorphism of the β_2-adrenoreceptor can alter the process of signal transduction by these receptors.[43,44] Three single-nucleotide polymorphisms in *ADRB2* have been associated with altered expression, down-regulation, or coupling of the receptor in response to β_2-adrenoreceptor agonists.[43] Single-nucleotide polymorphisms resulting in an Arg-to-Gly amino acid change at codon 16 and a Gln-to-Glu change at codon 27 are relatively common, with allele frequencies of 0.4 to 0.6, and are under intensive investigation for their clinical relevance.

A recent study of agonist-mediated vasodilatation and desensitization[44] revealed that patients who were homozygous for Arg at *ADRB2* codon 16 had nearly complete desensitization after continuous infusion of isoproterenol, with venodilatation decreasing from 44 percent at base line to 8 percent after 90 minutes of infusion (Fig. 6.4). In contrast,

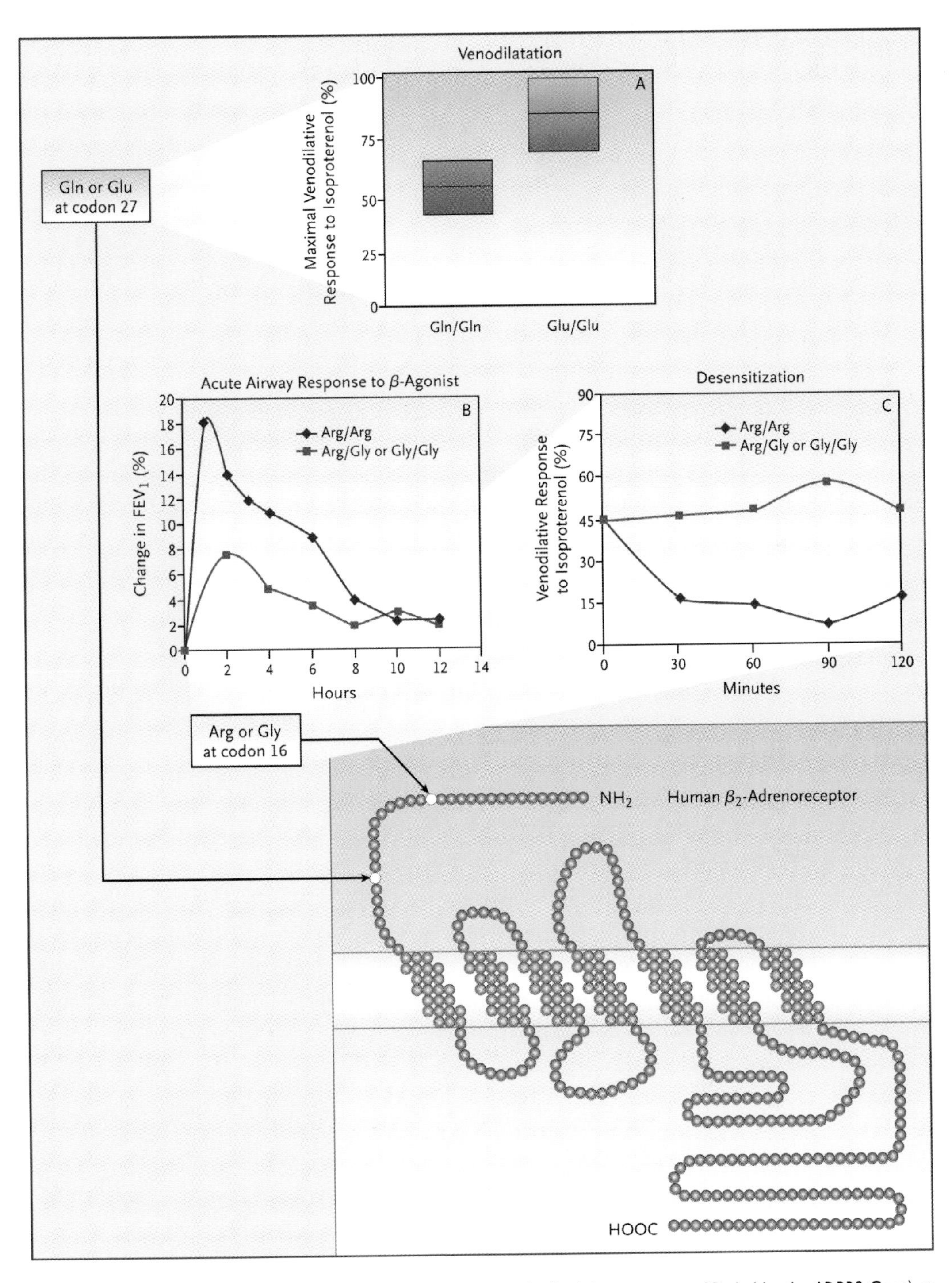

Figure 6.4 Functional Consequence of Genetic Polymorphisms in the β_2-Adrenoreceptor (Coded by the *ADRB2* Gene) at Codons 16 and 27. A homozygous Glu genotype at codon 27 is associated with greater venodilatation after the administration of isoproterenol (Panel A).[44] A homozygous Arg genotype at codon 16 is associated with greater airway response to oral albuterol (Panel B)[48] and greater desensitization to isoproterenol (Panel C).[44] FEV_1 denotes forced expiratory volume in one second.

patients homozygous for Gly at codon 16 had no significant change in venodilatation, regardless of their codon 27 status. Polymorphism at codon 27 was also of functional relevance; subjects homozygous for the Glu allele had higher maximal venodilatation in response to isoproterenol than those with the codon 27 Gln genotype, regardless of their codon 16 status (Fig. 6.4).[44]

These results are generally consistent with those of studies showing that the forced expiratory volume in one second (FEV_1) after a single oral dose of albuterol was higher by a factor of 6.5 in patients with the Arg/Arg genotype at codon 16 of *ADRB2* than in those with the Gly/Gly genotype (Fig. 6.4).[48] However, the influence of this genotype was different in patients receiving long-term, regularly scheduled therapy with inhaled β-agonists. Among these patients, those with the Arg/Arg genotype had a gradual decline in the morning peak expiratory flow measured before they had used medication, whereas no change was observed in patients with the Gly/Gly genotype.[47] In addition, the morning peak expiratory flow deteriorated dramatically after the cessation of therapy in patients with the Arg/Arg genotype, but not in those with the Gly/Gly genotype.[47] These data suggest that a codon 16 Arg/Arg genotype may identify patients at risk for deleterious or nonbeneficial effects of regularly scheduled therapy with inhaled β-agonists; the data also suggest that these patients may be candidates for alternative schedules of therapy, earlier initiation of anti-inflammatory agents, or both. These findings are also consistent with the aforementioned desensitization of the β_2-adrenoreceptor in patients with a codon 16 Arg/Arg genotype.[44]

At least 13 distinct single-nucleotide polymorphisms have been identified in *ADRB2*.[46] This finding has led to evaluation of the importance of haplotype structure as compared with individual single-nucleotide polymorphisms in determining receptor function and pharmacologic response. Among 77 white, black, Asian, and Hispanic subjects, only 12 distinct haplotypes of the 8192 possible *ADRB2* haplotypes were actually observed.[46] The bronchodilator response to inhaled β-agonist therapy in patients with asthma revealed a stronger association between bronchodilator response and haplotype than between bronchodilator response and any single-nucleotide polymorphism alone.[46] This is not surprising, because haplotype structure is often a better predictor of phenotypic consequences than are individual polymorphisms. This result suggests that it would be desirable to develop simple but robust molecular methods to determine the haplotype structure of patients.[68]

GENETIC POLYMORPHISMS WITH INDIRECT EFFECTS ON DRUG RESPONSE

Polymorphisms in genes encoding proteins that are neither direct targets of medications nor involved in their disposition have been shown to alter the response to treatment in certain situations (Table 6.2). For example, inherited differences in coagulation factors can predispose women taking oral contraceptives to deep-vein or cerebral-vein thrombosis,[80] whereas polymorphisms in the gene for the cholesterol ester transfer protein have been linked to the progression of atherosclerosis with pravastatin therapy.[75]

Genetic variation in cellular ion transporters can also have an indirect role in predisposing patients to toxic effects of drugs. For example, patients with variant alleles for sodium

Gene or Gene Product	Disease or Response Association	Medication	Influence of Polymorphism on Drug Effect or Toxicity
Adducin	Hypertension	Diuretics	Myocardial infarction or strokes[69]
Apolipoprotein E (APOE)	Progression of athero-sclerosis, ischemic cardio-vascular events	Statins (e.g., simvastatin)	Enhanced survival[70,71]
Apolipoprotein E (APOE)	Alzheimer's disease	Tacrine	Clinical improvement[72]
HLA	Toxicity	Abacavir	Hypersensitivity reaction[73,74]
Cholesterol ester transfer protein (CETP)	Progression of athero-sclerosis	Statins (e.g., pravastatin)	Slowing of progression of atherosclerosis by pravastatin[75]
Ion channels (HERG, KvLQT1, Mink, MiRP1)	Congenital long-QT syndrome	Erythromycin, terfen-adine,cisapride, cla-rithromycin, quinidine	Increased risk of drug-induced torsade de pointes[76–78]
Methylguanine methyl-transferase (MGMT)	Glioma	Carmustine	Response of glioma to carmustine[63]
Parkin	Parkinson's disease	Levodopa	Clinical improvement and levodopa-induced dyskinesias[79]
Prothrombin and factor V	Deep-vein thrombosis and cerebral-vein thrombosis	Oral contraceptives	Increased risk of deep-vein and cerebral-vein thrombosis with oral contraceptives[80]
Stromelysin-1	Atherosclerosis progression	Statins (e.g., pravastatin)	Reduction in cardiovascular events by pravastatin (death, myocardial infarction, stroke, angina, and others); reduction in risk of repeated angioplasty[81]

Table 6.2 Genetic Polymorphisms in Disease-Modifying or Treatment-Modifying Genes That Can Influence Drug Response*

*The examples shown are illustrative and not representative of all published studies, which exceed the scope of this review.

or potassium transporters may have substantial morbidity or mortality resulting from drug-induced long-QT syndrome. A mutation in *KCNE2*, the gene for an integral membrane subunit that assembles with HERG to form I_{Kr} potassium channels, was identified in a patient who had cardiac arrhythmia after receiving clarithromycin.[76] Additional *KCNE2* variants have been associated with the development of a very long QT interval after therapy with trimethoprim–sulfamethoxazole, with sulfamethoxazole inhibiting potassium channels encoded by the *KCNE2* (8T $\rightarrow$ A) variant.[77] Because *KCNE2* variants occur in about 1.6 percent of the population and their effect on drug actions can cause death, they are excellent candidates for polygenic strategies to prevent serious drug-induced toxic effects.

Genetic polymorphism in the apolipoprotein E (*APOE*) gene appears to have a role in predicting responses to therapy for Alzheimer's disease and to lipid-lowering drugs.[70,71,82,83] There are numerous allelic variants of the human *APOE* gene (e.g., *APOE* ϵ3, *APOE* ϵ4, *APOE* ϵ5, etc.), which contain one or more single-nucleotide polymorphisms that alter the amino acid sequence of the encoded protein (e.g., apolipoprotein ϵ4 has a Cys112Arg change). In a study of treatment of Alzheimer's disease with tacrine, 83 percent of the patients without any *APOE* ϵ4 allele showed improvement in total response and cognitive response after 30 weeks, as compared with 40 percent of patients with at least one *APOE* ϵ4 allele.[72] However, the greatest individual improvement in this study was seen in a patient with a single *APOE* ϵ4 allele, the unfavorable genotype, illustrating that a single gene will not always predict the response to a given treatment.[72] Follow-up studies indicate that the interaction between tacrine treatment and *APOE* genotype was strongest for women, again suggesting that many genes are involved in determining the efficacy of a treatment.[84]

The molecular basis for an association between apolipoprotein genotype and tacrine efficacy has not been elucidated, but it has been postulated that the *APOE* ϵ4 genotype may have an effect on cholinergic dysfunction in Alzheimer's disease that cannot be consistently overcome by therapy with acetylcholinesterase inhibitors such as tacrine. A randomized, placebo-controlled study of the noradrenergic vasopressinergic agonist S12024 in patients with Alzheimer's disease found the greatest protection of cognition in patients with the *APOE* ϵ4 genotype.[85] Confirmation of these results may offer an approach to the selection of initial therapy for Alzheimer's disease, with S12024 or similar medications being recommended for patients carrying an *APOE* ϵ4 allele.

Both phenotypic analysis and genotypic analysis of the *APOE* polymorphism have shown an association between *APOE* genotype and the response to lipid-lowering medications.[82,86–89] In most studies, patients with an *APOE* ϵ2 allele had the greatest diminution of low-density lipoprotein cholesterol after drug therapy. The decrease was greatest for those with *APOE* ϵ2, followed by *APOE* ϵ3 and then *APOE* ϵ4. This result was observed after treatment with a diverse range of lipid-lowering agents, including probucol, gemfibrozil, and many different 3-hydroxy-3-methylglutaryl-coenzyme A–reductase inhibitors (statins).[83] However, a significant effect of *APOE* genotype on the response to lipid-lowering agents has not been observed in all studies.[83]

In addition, although the *APOE4* allele was associated with less reduction in total and low-density lipoprotein cholesterol and a smaller increase in high-density lipoprotein cholesterol after fluvastatin therapy, there was no apparent influence of genotype on the progression of coronary artery disease or the incidence of clinical events.[88] Thus, prospective clinical evaluations with robust clinical end points and sufficient sample sizes are needed to define better the usefulness of the *APOE* genotype in selecting the treatment of hyperlipidemia and cardiovascular disease. The potential usefulness of the *APOE* genotype in predicting treatment response must be balanced by the concern that it could be used by insurance companies, health systems, and others to identify those at high risk for Alzheimer's disease, coronary artery disease, and possibly other illnesses.[82]

MOLECULAR DIAGNOSTIC METHODS FOR OPTIMIZING DRUG THERAPY

The potential is enormous for pharmacogenomics to yield a powerful set of molecular diagnostic methods that will become routine tools with which clinicians will select medications and drug doses for individual patients. A patient's genotype needs to be determined only once for any given gene, because except for rare somatic mutations, it does not change. Genotyping methods are improving so rapidly that it will soon be simple to test for thousands of single-nucleotide polymorphisms in one assay. It may be possible to collect a single blood sample from a patient, submit a small aliquot for analysis of a panel of genotypes (e.g., 20,000 single-nucleotide polymorphisms in 5000 genes), and test for those that are important determinants of drug disposition and effects. In our opinion, genotyping results will be of greatest clinical value if they are reported and interpreted according to the patient's diagnosis and recommended treatment options.

CHALLENGES FOR THE FUTURE

There are a number of critical issues that must be considered as strategies are developed to elucidate the inherited determinants of drug effects. A formidable one is that the inherited component of the response to drugs is often polygenic (Fig. 6.1). Approaches for elucidating polygenic determinants of drug response include the use of anonymous single-nucleotide polymorphism maps to perform genome-wide searches for polymorphisms associated with drug effects, and candidate-gene strategies based on existing knowledge of a medication's mechanisms of action and pathways of metabolism and disposition. Both these strategies have potential value and limitations, as shown in previous reviews.[5,90,91] However, the candidate-gene strategy has the advantage of focusing resources on a manageable number of genes and polymorphisms that are likely to be important, and it has produced encouraging results in a number of studies.[20,52] The limitations of this approach are the incompleteness of knowledge of a medication's pharmacokinetics and mechanisms of action. Gene-expression profiling[92,93] and proteomic studies[94] are evolving strategies for identifying genes that may influence drug response.

One of the most important challenges in defining pharmacogenetic traits is the need for well-characterized patients who have been uniformly treated and systematically evaluated to make it possible to quantitate drug response objectively. To this end, the norm should be to obtain genomic DNA from all patients enrolled in clinical drug trials, along with appropriate consent to permit pharmacogenetic studies. Because of marked population heterogeneity, a specific genotype may be important in determining the effects of a medication for one population or disease but not for another; therefore, pharmacogenomic relations must be validated for each therapeutic indication and in different racial and ethnic groups. Remaining cognizant of these caveats will help ensure accurate elucidation of genetic determinants of drug response and facilitate the translation of pharmacogenomics into widespread clinical practice.

Acknowledgments

Supported in part by grants from the National Institutes of Health (R37 CA36401, R01 CA78224, U01 GM61393, U01 GM61394, and U01 GM63340), Cancer Center support grants (CA21765 and CA091842), a Center of Excellence grant from the State of Tennessee, a grant from the Siteman Cancer Center, and a grant from American Lebanese Syrian Associated Charities.

Dr. Evans became a member of the Clinical Genomics Advisory Board of Merck and a member of the Scientific Advisory Board for Signature Genetics and Gentris after this review was written, and he was formerly a member of the Scientific Advisory Board of PPGX. He currently serves as a consultant to Bristol-Myers Squibb. He holds no equity positions in any of these companies. Dr. Evans's laboratory is supported by National Institutes of Health grants. He receives no research support from public or private companies. Dr. McLeod's laboratory is supported by grants from the National Institutes of Health, as well as by research grants from Novartis Pharmaceuticals and Ortho Clinical Diagnostics for projects that do not overlap directly or indirectly with the contents of this chapter.

References

1. Vesell ES. Pharmacogenetic perspectives gained from twin and family studies. Pharmacol Ther 1989; 41:535-52.

2. Kalow W, Tang BK, Endrenyi I. Hypothesis: comparisons of inter- and intra-individual variations can substitute for twin studies in drug research. Pharmacogenetics 1998; 8:283–9.

3. Evans WE, Relling MV. Pharmacogenomics: translating functional genomics into rational therapeutics. Science 1999; 286:487–91.

4. Evans WE, Johnson JA. Pharmacogenomics: the inherited basis for interindividual differences in drug response. Annu Rev Genomics Hum Genet 2001; 2:9–39.

5. McLeod HL, Evans WE. Pharmacogenomics: unlocking the human genome for better drug therapy. Annu Rev Pharmacol Toxicol 2001; 41:101–21.

6. Kalow W. Familial incidence of low pseudocholinesterase level. Lancet 1956; 2:576.

7. Carson PE, Flanagan CL, Ickes CE, Alving AS. Enzymatic deficiency in primaquine-sensitive erythrocytes. Science 1956; 124:484–5.

8. Hughes HB, Biehl JP, Jones AP, Schmidt LH. Metabolism of isoniazid in man as related to the occurrence of peripheral neuritis. Am Rev Tuberc 1954; 70:266–73.

9. Evans DAP, Manley KA, McKusick VA. Genetic control of isoniazid metabolism in man. Br Med J 1960; 2:485–91.

10. Sachidanandam R, Weissman D, Schmidt SC, et al. A map of human genome sequence variation containing 1.42 million single nucleotide polymorphisms. Nature 2001; 409:928–33.

11. Yates CR, Krynetski EY, Loennechen T, et al. Molecular diagnosis of thiopurine S-methyltransferase deficiency: genetic basis for azathioprine and mercaptopurine intolerance. Ann Intern Med 1997; 126:608–14.

12. Weinshilboum R. Inheritance and drug response. N Engl J Med 2003; 348:529–37.

13. Meyer UA. Pharmacogenetics and adverse drug reactions. Lancet 2000; 356:1667–71.

14. Ingelman-Sundberg M, Oscarson M, McLellan RA. Polymorphic human cytochrome P450 enzymes: an opportunity for individualized drug treatment. Trends Pharmacol Sci 1999; 20:342–9.

15. Kuehl P, Zhang J, Lin Y, et al. Sequence diversity in CYP3A promoters and characterization of the genetic basis for polymorphic CYP3A5 expression. Nat Genet 2001; 27:383–91.

16. Sata F, Sapone A, Elizondo G, et al. CYP3A4 allelic variants with amino acid substitutions in exons 7 and 12: evidence for an allelic variant with altered catalytic activity. Clin Pharmacol Ther 2000; 67:48–56.

17. Borst P, Evers R, Kool M, Wijnholds J. A family of drug transporters: the multidrug resistance-associated proteins. J Natl Cancer Inst 2000; 92: 1295–302.

18. Kim RB, Leake BF, Choo EF, et al. Identification of functionally variant MDR1 alleles among European Americans and African Americans. Clin Pharmacol Ther 2001; 70:189–99.

19. Hoffmeyer S, Burk O, von Richter O, et al. Functional polymorphisms of the human multidrug-resistance gene: multiple sequence variations and correlation of one allele with P-glycoprotein expression and activity in vivo. Proc Natl Acad Sci U S A 2000; 97:3473–8.

20. Fellay J, Marzolini C, Meaden ER, et al. Response to antiretroviral treatment in HIV-1-infected individuals with allelic variants of the multidrug resistance transporter 1: a pharmacogenetics study. Lancet 2002; 359:30–6.

21. Choo EF, Leake B, Wandel C, et al. Pharmacological inhibition of P-glycoprotein transport enhances the distribution of HIV-1 protease

inhibitors into brain and testes. Drug Metab Dispos 2000; 28:655–60.

22. Brinkmann U, Roots I, Eichelbaum M. Pharmacogenetics of the human drug-transporter gene MDR1: impact of polymorphisms on pharmacotherapy. Drug Discov Today 2001; 6:835–9.

23. Rao VV, Dahlheimer JL, Bardgett ME, et al. Choroid plexus epithelial expression of MDR1 P glycoprotein and multidrug resistance-associated protein contribute to the blood-cerebrospinal-fluid drug-permeability barrier. Proc Natl Acad Sci U S A 1999; 96:3900–5.

24. Thiebaut F, Tsuruo T, Hamada H, Gottesman MM, Pastan I, Willingham MC. Cellular localization of the multidrug-resistance gene product P-glycoprotein in normal human tissues. Proc Natl Acad Sci U S A 1987; 84:7735–8.

25. Schinkel AH, Wagenaar E, Mol CA, van Deemter L. P-glycoprotein in the blood-brain barrier of mice influences the brain penetration and pharmacological activity of many drugs. J Clin Invest 1996; 97:2517–24.

26. Hitzl M, Drescher S, van der Kuip H, et al. The C3435T mutation in the human MDR1 gene is associated with altered efflux of the P-glycoprotein substrate rhodamine 123 from CD56+ natural killer cells. Pharmacogenetics 2001; 11:293–8.

27. Sakaeda T, Nakamura T, Horinouchi M, et al. MDR1 genotype-related pharmacokinetics of digoxin after single oral administration in healthy Japanese subjects. Pharm Res 2001; 18:1400–4.

28. Ameyaw MM, Regateiro F, Li T, et al. MDR1 pharmacogenetics: frequency of the C3435T mutation in exon 26 is significantly influenced by ethnicity. Pharmacogenetics 2001; 11:217–21.

29. McLeod H. Pharmacokinetic differences between ethnic groups. Lancet 2002; 359:78.

30. Schaeffeler E, Eichelbaum M, Brinkmann U, et al. Frequency of C3435T polymorphism of MDR1 gene in African people. Lancet 2001; 358:383–4.

31. Schuetz JD, Connelly MC, Sun D, et al. MRP4: a previously unidentified factor in resistance to nucleoside-based antiviral drugs. Nat Med 1999; 5:1048–51.

32. Jacobsen P, Rossing K, Rossing P, et al. Angiotensin converting enzyme gene polymorphism and ACE inhibition in diabetic nephropathy. Kidney Int 1998; 53:1002–6.

33. Kohno M, Yokokawa K, Minami M, et al. Association between angiotensin-converting enzyme gene polymorphisms and regression of left ventricular hypertrophy in patients treated with angiotensin-converting enzyme inhibitors. Am J Med 1999; 106: 544–9.

34. Ohmichi N, Iwai N, Uchida Y, Shichiri G, Nakamura Y, Kinoshita M. Relationship between the response to the angiotensin converting enzyme inhibitor imidapril and the angiotensin converting enzyme genotype. Am J Hypertens 1997; 10:951–5.

35. Okamura A, Ohishi M, Rakugi H, et al. Pharmacogenetic analysis of the effect of angiotensin-converting enzyme inhibitor on restenosis after percutaneous transluminal coronary angioplasty. Angiology 1999; 50:811–22.

36. Penno G, Chaturvedi N, Talmud PJ, et al. Effect of angiotensin-converting enzyme (ACE) gene polymorphism on progression of renal disease and the influence of ACE inhibition in IDDM patients: findings from the EUCLID Randomized Controlled Trial: EURODIAB Controlled Trial of Lisinopril in IDDM. Diabetes 1998; 47:1507–11.

37. Perna A, Ruggenenti P, Testa A, et al. ACE genotype and ACE inhibitors induced renoprotection in chronic proteinuric nephropathies. Kidney Int 2000; 57:274–81.

38. Prasad A, Narayanan S, Husain S, et al. Insertion-deletion polymorphism of the ACE gene modulates reversibility of endothelial dysfunction with ACE inhibition. Circulation 2000; 102:35–41.

39. Sasaki M, Oki T, Iuchi A, et al. Relationship between the angiotensin converting enzyme gene polymorphism and the effects of enalapril on left ventricular hypertrophy and impaired diastolic filling in essential hypertension: M-mode and pulsed Doppler echocardiographic studies. J Hypertens 1996; 14:1403–8.

40. Stavroulakis GA, Makris TK, Krespi PG, et al. Predicting response to chronic antihypertensive treatment with fosinopril: the role of angiotensin-converting enzyme gene polymorphism. Cardiovasc Drugs Ther 2000; 14:427–32.

41. Marian AJ, Safavi F, Ferlic L, Dunn JK, Gotto AM, Ballantyne CM. Interactions between angiotensin-I converting enzyme insertion/deletion polymorphism and response of plasma lipids and coronary atherosclerosis to treatment with fluvastatin: the Lipoprotein and Coronary Atherosclerosis Study. J Am Coll Cardiol 2000; 35:89–95.

42. Drazen JM, Yandava CN, Dube L, et al. Pharmacogenetic association between ALOX5 promoter genotype and the response to anti-asthma treatment. Nat Genet 1999; 22:168–70.

43. Liggett SB. Beta(2)-adrenergic receptor pharmacogenetics. Am J Respir Crit Care Med 2000; 161:S197–S201.

44. Dishy V, Sofowora GG, Xie H-G, et al. The effect of common polymorphisms of the β_2-adrenergic receptor on agonist-mediated vascular desensitization. N Engl J Med 2001; 345:1030–5.

45. Cockcroft JR, Gazis AG, Cross DJ, et al. Beta(2)-adrenoceptor polymorphism determines vascular reactivity in humans. Hypertension 2000; 36:371–5.

46. Drysdale CM, McGraw DW, Stack CB, et al. Complex promoter and coding region beta 2-adrenergic receptor haplotypes alter receptor expression and predict in vivo responsiveness. Proc Natl Acad Sci U S A 2000; 97:10483–8.

47. Israel E, Drazen JM, Liggett SB, et al. Effect of polymorphism of the beta(2)-adrenergic receptor on response to regular use of albuterol in asthma. Int Arch Allergy Immunol 2001; 124:183–6.

48. Lima JJ, Thomason DB, Mohamed MH, Eberle LV, Self TH, Johnson JA. Impact of genetic polymorphisms of the beta2-adrenergic receptor on albuterol bronchodilator pharmacodynamics. Clin Pharmacol Ther 1999; 65:519–25.

49. Martinez FD, Graves PE, Baldini M, Solomon S, Erickson R. Association between genetic polymorphisms of the beta2-adrenoceptor and response to albuterol in children with and without a history of wheezing. J Clin Invest 1997; 100:3184–8.

50. Tan S, Hall IP, Dewar J, Dow E, Lipworth B. Association between beta 2-adrenoceptor polymorphism and susceptibility to bronchodilator desensitization in moderately severe stable asthmatics. Lancet 1997; 350:995–9.

51. Mukae S, Aoki S, Itoh S, Iwata T, Ueda H, Katagiri T. Bradykinin B(2) receptor gene polymorphism is associated with angiotensin-converting enzyme inhibitor-related cough. Hypertension 2000; 36:127–31.

52. Arranz MJ, Munro J, Birkett J, et al. Pharmacogenetic prediction of clozapine response. Lancet 2000; 355:1615–6.

53. Basile VS, Masellis M, Badri F, et al. Association of the MscI polymorphism of the dopamine D3 receptor gene with tardive dyskinesia in schizophrenia. Neuropsychopharmacology 1999; 21:17–27.

54. Eichhammer P, Albus M, Borrmann-Hassenbach M, et al. Association of dopamine D3-receptor gene variants with neuroleptic induced akathisia in schizophrenic patients: a generalization of Steen's study on DRD3 and tardive dyskinesia. Am J Med Genet 2000; 96:187–91.

55. Hwu HG, Hong CJ, Lee YL, Lee PC, Lee SF. Dopamine D4 receptor gene polymorphisms and neuroleptic response in schizophrenia. Biol Psychiatry 1998; 44:483–7.

56. Kaiser R, Konneker M, Henneken M, et al. Dopamine D4 receptor 48-bp repeat polymorphism: no association with response to antipsychotic treatment, but association with catatonic schizophrenia. Mol Psychiatry 2000; 5:418–24.

57. Ongphiphadhanakul B, Chanprasertyothin S, Payatikul P, et al. Oestrogen-receptor-alpha gene polymorphism affects response in bone mineral density to oestrogen in post-menopausal women. Clin Endocrinol (Oxf) 2000; 52:581–5.

58. Herrington DM, Howard TD, Hawkins GA, et al. Estrogen-receptor polymorphisms and effects of estrogen replacement on high-density lipoprotein cholesterol in women with coronary disease. N Engl J Med 2002; 346:967–74.

59. Michelson AD, Furman MI, Goldschmidt-Clermont P, et al. Platelet GP IIIa PI(A) polymorphisms display different sensitivities to agonists. Circulation 2000; 101:1013–8.

60. Kim DK, Lim SW, Lee S, et al. Serotonin transporter gene polymorphism and antidepressant response. Neuroreport 2000; 11:215–9.

61. Smeraldi E, Zanardi R, Benedetti F, Di Bella D, Perez J, Catalano M. Polymorphism within the promoter of the serotonin transporter gene and antidepressant efficacy of fluvoxamine. Mol Psychiatry 1998; 3:508–11.

62. Whale R, Quested DJ, Laver D, Harrison PJ, Cowen PJ. Serotonin transporter (5-HTT) promoter genotype may influence the prolactin response to clomipramine. Psychopharmacology (Berl) 2000; 150:120–2.

63. Esteller M, Garcia-Foncillas J, Andion E, et al. Inactivation of the DNA-repair gene MGMT and the clinical response of gliomas to alkylating agents. N Engl J Med 2000; 343:1350–4. [Erratum, N Engl J Med 2000; 343:1740.]

64. Evans WE, Hon YY, Bomgaars L, et al. Preponderance of thiopurine S-methyltransferase deficiency and heterozygosity among patients intolerant to mercaptopurine or azathioprine. J Clin Oncol 2001; 19:2293–301.

65. Black AJ, McLeod HL, Capell HA, et al. Thiopurine methyltransferase genotype predicts therapy-limiting severe toxicity from azathioprine. Ann Intern Med 1998; 129:716–8.

66. Relling MV, Hancock ML, Rivera GK, et al. Mercaptopurine therapy intolerance and heterozygosity at the thiopurine S-methyltransferase gene locus. J Natl Cancer Inst 1999; 91:2001–8.

67. Relling MV, Rubnitz JE, Rivera GK, et al. High incidence of secondary brain tumours after radiotherapy and antimetabolites. Lancet 1999; 354:34–9.

68. McDonald OG, Krynetski EY, Evans WE. Molecular haplotyping of genomic DNA for multiple single-nucleotide polymorphisms located kilobases apart using long-range polymerase chain reaction and intramolecular ligation. Pharmacogenetics 2002; 12:93–9.

69. Psaty BM, Smith NL, Heckbert SR, et al. Diuretic therapy, the α-adducin gene variation, and the risk of myocardial infarction or stroke in persons with treated hypertension. JAMA 2002; 287:1680–9.

70. Gerdes LU, Gerdes C, Kervinen K, et al. The apolipoprotein epsilon4 allele determines prognosis and the effect on prognosis of simvastatin in survivors of myocardial infarction: a substudy of the Scandinavian Simvastatin Survival Study. Circulation 2000; 101:1366–71.

71. Ordovas JM, Lopez-Miranda J, Perez-Jimenez F, et al. Effect of apolipoprotein E and A-IV phenotypes on the low density lipoprotein response to HMG CoA reductase inhibitor therapy. Atherosclerosis 1995; 113:157–66.

72. Poirier J, Delisle MC, Quirion R, et al. Apolipoprotein E4 allele as a predictor of cholinergic deficits and treatment outcome in Alzheimer disease. Proc Natl Acad Sci U S A 1995; 92:12260–4.

73. Mallal S, Nolan D, Witt C, et al. Association between presence of HLA-B*5701, HLA-DR7, and HLA-DQ3 and hypersensitivity to HIV-1 reverse-transcriptase inhibitor abacavir. Lancet 2002; 359: 727–32.

74. Hetherington S, Hughes AR, Mosteller M, et al. Genetic variations in HLA-B region and hypersensitivity reaction to abacavir. Lancet 2002; 359:1121–2.

75. Kuivenhoven JA, Jukema JW, Zwinderman AH, et al. The role of a common variant of the cholesteryl ester transfer protein gene in the progression of coronary atherosclerosis. N Engl J Med 1998; 338:86–93.

76. Abbott GW, Sesti F, Splawski I, et al. MiRP1 forms IKr potassium channels with HERG and is associated with cardiac arrhythmia. Cell 1999; 97:175–87.

77. Sesti F, Abbott GW, Wei J, et al. A common polymorphism associated with antibiotic-induced cardiac arrhythmia. Proc Natl Acad Sci U S A 2000; 97: 10613–8.

78. Napolitano C, Schwartz PJ, Brown AM, et al. Evidence for a cardiac ion channel mutation underlying drug-induced QT prolongation and life-threatening arrhythmias. J Cardiovasc Electrophysiol 2000; 11:691–6.

79. Lücking CB, Dürr A, Bonifati V, et al. Association between early-onset Parkinson's disease and mutations in the *parkin* gene. N Engl J Med 2000; 342:1560–7.

80. Martinelli I, Sacchi E, Landi G, Taioli E, Duca F, Mannucci PM. High risk of cerebral-vein thrombosis in carriers of a prothrombin-gene mutation and in users of oral contraceptives. N Engl J Med 1998; 338:1793–7.

81. de Maat MP, Jukema JW, Ye S, et al. Effect of the stromelysin-1 promoter on efficacy of pravastatin in coronary atherosclerosis and restenosis. Am J Cardiol 1999; 83:852–6.

82. Issa AM, Keyserlingk EW. Apolipoprotein E genotyping for pharmacogenetic purposes in Alzheimer's disease: emerging ethical issues. Can J Psychiatry 2000; 45:917–22.

83. Siest G, Bertrand P, Herbeth B, et al. Apolipoprotein E polymorphisms and concentration in chronic diseases and drug responses. Clin Chem Lab Med 2000; 38:841–52.

84. Farlow MR, Lahiri DK, Poirier J, Davignon J, Schneider L, Hui SL. Treatment outcome of tacrine therapy depends on apolipoprotein genotype and gender of the subjects with Alzheimer's disease. Neurology 1998; 50:669–77.

85. Richard F, Helbecque N, Neuman E, Guez D, Levy R, Amouyel P. APOE genotyping and response to drug treatment in Alzheimer's disease. Lancet 1997; 349:539.

86. Nestruck AC, Bouthillier D, Sing CF, Davignon J. Apolipoprotein E polymorphism and plasma cholesterol response to probucol. Metabolism 1987; 36:743–7.

87. Pedro-Botet J, Schaefer EJ, Bakker-Arkema RG, et al. Apolipoprotein E genotype affects plasma lipid response to atorvastatin in a gender specific manner. Atherosclerosis 2001; 158:183–93.

88. Watanabe J, Kobayashi K, Umeda F, et al. Apolipoprotein E polymorphism affects the response to pravastatin on plasma apolipoproteins in diabetic patients. Diabetes Res Clin Pract 1993; 20:21–7.

89. Ballantyne CM, Herd JA, Stein EA, et al. Apolipoprotein E genotypes and response of plasma lipids and progression-regression of coronary atherosclerosis to lipid-lowering drug therapy. J Am Coll Cardiol 2000; 36:1572–8.

90. Cargill M, Daley GQ. Mining for SNPs: putting the common variants–common disease hypothesis to the test. Pharmacogenomics 2000; 1:27–37.

91. Sham P. Shifting paradigms in gene-mapping methodology for complex traits. Pharmacogenomics 2001; 2:195–202.

92. Staunton JE, Slonim DK, Coller HA, et al. Chemosensitivity prediction by transcriptional profiling. Proc Natl Acad Sci U S A 2001; 98:10787–92.

93. Yeoh EJ, Ross ME, Shurtleff SA, et al. Classification, subtype discovery, and prediction of outcome in pediatric acute lymphoblastic leukemia by gene expression profiling. Cancer Cell 2002; 1:133–43.

94. Liotta LA, Kohn EC, Petricoin EF. Clinical proteomics: personalized molecular medicine. JAMA 2001; 286:2211–4.

7

Pharmacogenetics in the Laboratory and the Clinic

DAVID B. GOLDSTEIN, Ph.D.

One of the most striking features of modern medicines is how often they fail to work. Even when they do work, they are often associated with serious adverse reactions. Indeed, adverse reactions to drugs rank as one of the leading causes of death and illness in the developed world.[1] How can we improve the success rate?

The Human Genome Project and other advances have generated expectations that medicines can be customized to match the genetic makeup of patients, thereby dramatically improving efficacy and safety. Although the prospects for basic research in pharmacogenetics look very promising, the incorporation into clinical practice of the data it generates presents considerable challenges.

Polymorphism of the genes encoding drug-metabolizing enzymes has been the core focus of pharmacogenetics since its beginnings in the 1950s, but other classes of genes influencing responses to drugs are receiving increasing attention.[2,3] In addition to the many well-known variants of drug-metabolizing enzymes,[2] pharmacogenetic variants have now been documented in drug transporters and targets, as well as in their associated pathways.[3]

Many of the genes that have been studied have variants that occur with moderate-to-high frequency and are known to influence the response to drugs or appear likely to do so (for example, because they change the activity of a relevant enzyme). These findings are important not only because of the specific variants that have been identified, but also because they suggest that it will be possible to find many of the variants that we do not yet know about. Much of the excitement about personalizing medicines is fueled by association studies, which seek to relate genetic variants to drug response. Genetic association studies will be more successful in finding common variants than rare variants, unless the variant has very high penetrance.

Originally published February 6, 2003

In a case–control study, genotypes of persons with and without a particular response of interest—either a therapeutic response or an adverse event—are compared. The key question is which genetic variants to consider. There are approximately 10 million single-nucleotide polymorphisms, or SNPs (variants with frequencies of more than 1 percent), in the human genome, of which as many as 4 million have been catalogued.[4] Because our knowledge in this area is incomplete and because of cost constraints, it is not possible to evaluate all polymorphic SNPs in the genome directly through association studies. Instead, the aim is to use associations among variants, or linkage disequilibrium,[5] to leverage the information provided by typing a subgroup of polymorphic markers.

Current thinking about the design of association studies has been greatly influenced by data showing that linkage disequilibrium in humans is composed of regions of low diversity of haplotypes, where the common haplotypes can be represented by knowledge of only a few genotypes in the block, and stretches of more rapid breakdown of linkage disequilibrium, where a few sentinel SNPs cannot reliably represent haplotype diversity; in many cases, the latter regions correspond to hot spots of meiotic recombination.[6,7] This structure of linkage disequilibrium inspired the idea of haplotype tagging, in which a set of SNPs is identified that tags each of the common haplotypes within a block of linkage disequilibrium. The general algorithm is straightforward: determine the haplotype structures of genes or genomic regions of interest in controls, identify tagging SNPs, and analyze the tagging SNPs in patients whose drug-response phenotype is known. By some estimates, an average of only five to seven SNPs per gene would be required to represent all the common polymorphisms in candidate genes. This means that it is now economical to conduct exhaustive studies of candidate genes once the haplotype structure of the genes has been determined.

It must be understood, however, that the tagging SNPs must be carefully selected and validated—not just any five to seven SNPs will do. Thus, many previous pharmacogenetic association studies have been incomplete, in that they did not fully represent the common variation in the genes studied, even when multiple polymorphisms within a gene were considered. For clarity, this approach should be referred to as the study of "candidate polymorphisms," to distinguish it from a systematic study of candidate genes using tagging SNPs for each common haplotype. Failure of a candidate-polymorphism study does not rule out involvement of the gene or genes, whereas a negative candidate-gene study can provide a statistical limit to the importance of any common variant in the gene. Given the economy provided by tagging SNPs, there is little justification now for not conducting a candidate-gene study, especially since the haplotype structures of genes need only be determined once in the population of interest. The tags can then be used for any clinical study. It should be noted, however, that haplotype tagging will often fail to capture rare variants, which are also known to be important in variable responses to a drug.[8]

There are several implications of this approach, termed "haplotype mapping," that are worth emphasizing. Limited diversity of the haplotypes of genes increases the prospects of identifying interactions between the haplotypes found in different genes—for example, elements of a common pathway. This opportunity argues strongly that potentially interacting genes are best analyzed as integrated sets (e.g., the genes encoding a receptor complex, the renin–angiotensin–aldosterone pathway, or drug-metabolizing enzymes).

Systematic association studies will very soon be feasible for large sets of candidate genes (i.e., sets numbering in the hundreds) within the context of academic laboratories, but systematic genome-wide analyses will not be possible for quite some time. The National Institutes of Health is spearheading a global effort to find an appropriate set of tagging SNPs for the entire genome and estimates that it will require 300,000 to 600,000 SNPs.[9] If it takes 450,000 SNPs at the (currently reasonable) cost of 10 cents per sample per genotype, the cost of analyzing 1000 cases and 1000 controls will be $90 million. On the other hand, a systematic candidate-gene study of the key elements of the renin–angiotensin pathway would cost several hundred thousand dollars at most, once the haplotype structure of the genes had been determined.

In short, much of the enthusiasm surrounding pharmacogenetics appears to be justified. Haplotype mapping is a powerful framework for finding common variants that influence drug response, and there is clear evidence that common variants play an important part in variable drug response. It does not follow, however, that pharmacogenetics will revolutionize health care overnight. The pharmacogenetics of asthma illustrates the reasons both for optimism about progress in research and for caution about translation into clinical practice. The β_2-adrenergic receptor is the target of the most commonly used medicines for asthma and is therefore a natural candidate for influencing the response to inhaled β-agonists. Indeed, the Arg–Arg genotype at position 16 in the β_2-adrenergic–receptor gene has been positively associated with the acute response to treatment.[10] Dynamic analyses, however, revealed that the Arg–Arg genotype is also associated with a significant decrease in response after regular use of β-agonists, whereas the Gly–Gly genotype was unaffected by regular use.[11] Combining these data with functional analyses, Liggett[12] explained the results with a model of receptor regulation, but the model has not been verified and the clinical implications remain unclear,[11] despite the extremely effective combination of association and functional data.

Similarly, P-glycoprotein is a nonspecific drug-efflux pump that is active at the blood–brain barrier and is known to act on antiepileptic drugs. My colleagues and I showed that a previously known high-activity variant in exon 26 of the encoding *ABCB1* gene is significantly associated with resistance to drug treatment in patients with epilepsy (unpublished data), suggesting the possibility that inhibition of P-glycoprotein might improve the response to treatment in some patients with refractory epilepsy. Previous clinical experience with P-glycoprotein inhibitors, however, provides a strong cautionary note concerning the clinical exploitation of this result. Clinical trials of P-glycoprotein inhibitors as a means of reducing the frequency of multidrug resistance in patients undergoing antitumor therapy have been generally disappointing, because of inadequate knowledge about the relevance of P-glycoprotein in individual patients, drug–drug interactions, the importance of other transporters, and perhaps unaccounted-for variation in the *ABCB1* gene itself.[13]

These and other examples suggest that, with only rare exceptions, the translation of pharmacogenetic research into clinical practice will be intellectually challenging, time-consuming, and expensive. It will usually require explicit clinical evaluation, meaning that translation will lag years behind basic research. There are steps, however, that the research community can take to improve the efficiency of translation.

Reported associations between genetic variants and drug responses should be as secure as possible, and every effort should be made to determine the causal variant or variants. It would be helpful to define minimal standards for reported associations. All reports should include an assessment of the structure of linkage disequilibrium surrounding the associated polymorphism in order to delimit an associated interval—that is, the boundaries on either side of the associated polymorphism that delimit an area of sufficiently high linkage disequilibrium that the causal variant responsible for the observed association could reside within it. In addition, all reports should include a check and, if necessary, correction for stratification of the population,[14] which can create spurious association, in order to reduce the amount of effort wasted on spurious associations.

Finally, the biologic function of causal variants must be assessed in order to aid in the interpretation of the associations. Particular care would be required in the use of diagnostic associations in the absence of identified causal variants, in part because patterns of linkage disequilibrium will differ among populations, thereby changing the associations between causal variants and the associated markers. Basic research in pharmacogenetics deserves the support and the excitement that it has generated, but this excitement should not lead to unrealistic expectations about the rate at which medicines can be personalized according to genotype.

References

1. Lazarou J, Pomeranz BH, Corey PN. Incidence of adverse drug reactions in hospitalized patients: a meta-analysis of prospective studies. JAMA 1998; 279:1200–5.
2. Weinshilboum R. Inheritance and drug response. N Engl J Med 2003; 348:529–37.
3. Evans WE, McLeod HL. Pharmacogenomics—drug disposition, drug targets, and side effects. N Engl J Med 2003; 348:538–49.
4. Gabriel SB, Schaffner SF, Nguyen H, et al. The structure of haplotype blocks in the human genome. Science 2002; 296:2225–9.
5. Guttmacher AE, Collins FS. Genomic medicine—a primer. N Engl J Med 2002; 347:1512–20.
6. Daly MJ, Rioux JD, Schaffner SF, Hudson TJ, Lander ES. High-resolution haplotype structure in the human genome. Nat Genet 2001; 29:229–32.
7. Jeffreys AJ, Kauppi L, Neumann R. Intensely punctate meiotic recombination in the class II region of the major histocompatibility complex. Nat Genet 2001; 29:217–22.
8. Relling MV, Dervieux T. Pharmacogenetics and cancer therapy. Nat Rev Cancer 2001; 1:99–108.
9. National Human Genome Research Institute newsroom. Bethesda, Md.: NHGRI, 2003. (Accessed January 10, 2003, at http://www.genome.gov.)
10. Martinez FD, Graves PE, Baldini M, Solomon S, Erickson R. Association between genetic polymorphisms of the beta2–adrenoceptor and the response to albuterol in children with and without a history of wheezing. J Clin Invest 1997; 100:3184–8.
11. Palmer LJ, Silverman ES, Weiss ST, Drazen JM. Pharmacogenetics of asthma. Am J Respir Crit Care Med 2002; 165:861–6.
12. Liggett SB. Pharmacogenetics of relevant targets in asthma. Clin Exp Allergy 1998; 28:Suppl:77–9.
13. Krishna R, Mayer LD. Multidrug resistance (MDR) in cancer: mechanisms, reversal using modulators of MDR and the role of MDR modulators in influencing the pharmacokinetics of anticancer drugs. Eur J Pharm Sci 2000; 11:265–83.
14. Reich DE, Goldstein DB. Detecting association in a case-control study while correcting for population stratification. Genet Epidemiol 2001; 20:4–16.

8

Hereditary Colorectal Cancer

HENRY T. LYNCH, M.D., and ALBERT DE LA CHAPELLE, M.D., Ph.D.

The annual incidence of colorectal cancer in the United States is approximately 148,300 (affecting 72,600 males and 75,700 females), with 56,600 deaths (in 27,800 males and 28,800 females).[1] The lifetime risk of colorectal cancer in the general population is about 5 to 6 percent.[1] Patients with a familial risk—those who have two or more first- or second-degree relatives (or both) with colorectal cancer—make up approximately 20 percent of all patients with colorectal cancer, whereas approximately 5 to 10 percent of the total annual burden of colorectal cancer is mendelian in nature—that is, it is inherited in an autosomal dominant manner. In this chapter we will focus on the two major forms of hereditary colorectal cancer, familial adenomatous polyposis and hereditary nonpolyposis colorectal cancer.

OVERALL CLINICAL APPROACH

The most important step leading to the diagnosis of a hereditary cancer syndrome is the compilation of a thorough family history of cancer.[2–4] A patient and his or her key relatives, working either alone or with a trained nurse or genetic counselor, can compile such a detailed family history. The focus should be on identifying cancer of all types and sites; the family member's age at the onset of cancer; any pattern of multiple primary cancers; any association with phenotypic features that may be related to cancer, such as colonic adenomas; and documentation of pathological findings whenever possible. This information will frequently identify a hereditary colorectal cancer syndrome in the family, should it exist. Molecular genetic testing may then provide verification of the diagnosis, when a germ-line mutation is present in the family.[5,6] The primary care physician may wish to refer the patient to a hereditary-cancer specialist and genetic counselor for further

Originally published March 6, 2003

evaluation should there be any remaining question about the disorder's clinical or molecular genetic diagnosis and the need for targeted surveillance and management.

Once a diagnosis of a hereditary colorectal cancer syndrome is established, the proband's high-risk relatives should be notified, and genetic counseling and DNA testing should be performed in consenting relatives, when such testing is appropriate. In an attempt to reduce morbidity and mortality, surveillance measures may then be instituted that reflect the natural history of the disorder.[7]

Once it is clear that a patient has a familial form of colorectal cancer, genetic counseling is mandatory and must provide the patient and his or her extended family with important details about their genetic risk of cancer at specific sites, on the basis of the natural history of the hereditary cancer syndrome; the options for surveillance and management; and the availability of genetic testing.[8,9] Counseling should be face to face, but a session may include multiple family members.[8] The concept of informed consent implies that a patient has received counseling, information, and putative test results and has signed a document to that effect. The results of tests for mutations should be revealed to the patient on a one-to-one basis.[7]

DIAGNOSTIC CLUES

Syndromes with distinguishing phenotypes, such as florid colonic adenomas in familial adenomatous polyposis, are easier to diagnose than hereditary disorders that lack clear phenotypic characteristics. For instance, the attenuated polyposis phenotype of familial adenomatous polyposis is characterized by a paucity of colonic adenomas, and the ones that do occur are primarily in the proximal colon. The onset of colorectal cancer is at a later average age (approximately 55 years) than that of classic familial adenomatous polyposis (approximately 39 years). These differences make it more difficult for clinicians to diagnose than its classic counterpart, despite their having a high index of suspicion for a familial colorectal cancer syndrome.[10,11]

In the case of hereditary nonpolyposis colorectal cancer, five cardinal features will help to identify affected families. The first is an earlier average age at the onset of cancer than in the general population; for example, the average age at the onset of hereditary nonpolyposis colorectal cancer is approximately 45 years,[7] whereas the average age at the onset of sporadic cases is approximately 63 years. The second feature is a particular pattern of primary cancers segregating within the pedigree, such as colonic and endometrial cancer.[7,12] The third is survival that differs from the norm for the specific cancer.[13–16] The fourth is distinguishing pathological features,[17,18] and the fifth and sine qua non is the identification of a germ-line mutation in affected members of the family.[5]

There are two broad classes of hereditary colorectal cancer, based on the predominant location of the cancer: distal and proximal. Colorectal cancers involving the distal colon are more likely to have aneuploid DNA, harbor mutations in the adenomatous polyposis coli *(APC)*, *p53*, and *K-ras* genes, and behave more aggressively[7]; proximal colorectal cancers are more likely to have diploid DNA, possess microsatellite instability, harbor mutations in the mismatch-repair genes, and behave less aggressively, as in hereditary nonpolyposis colorectal

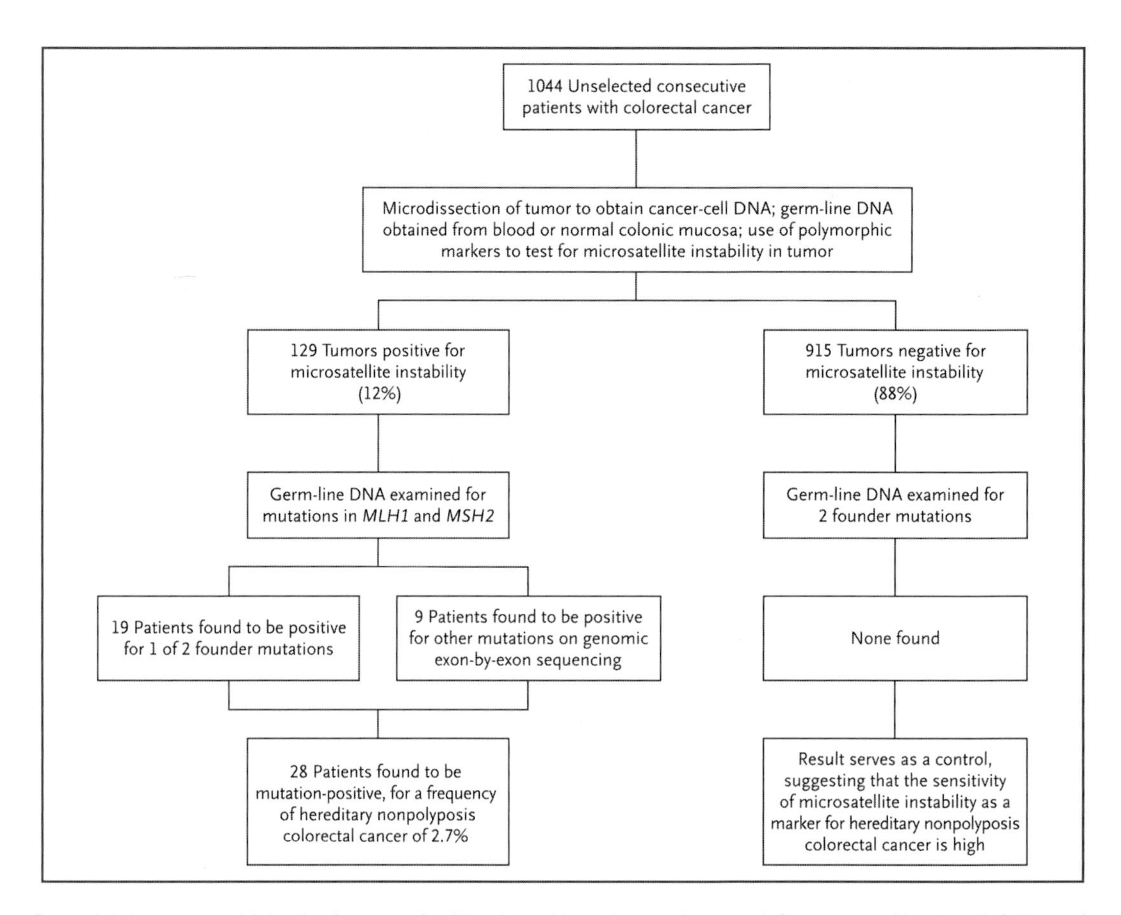

Figure 8.1 Approach to Molecular Screening for Hereditary Nonpolyposis Colorectal Cancer in an Unselected Cohort of Consecutive Patients with Newly Diagnosed Colorectal Cancer. Data are from Aaltonen et al.[22] and Salovaara et al.[23] This screening strategy relies on microsatellite instability as a primary marker for hereditary nonpolyposis colorectal cancer.[19,24] In most studies of unselected patients with colorectal cancers, the proportion who are positive for microsatellite instability ranges from 12 to 16 percent.[25] For this purpose, microsatellite instability can be determined with the use of just one or two markers and, in many cases, without the need for matching normal DNA.[26,27] Fixed, paraffin-embedded tumor specimens are a readily available source of DNA for this test, but the specimen must be determined histologically to contain at least 30 to 50 percent tumor cells. As a source of germ-line DNA for the detection of mutations, a blood sample is most suitable. The proportion of all patients with colorectal cancer who have hereditary nonpolyposis colorectal cancer may vary among populations. The proportion found in these studies (2.7 percent) is an underestimate, because neither microsatellite-instability testing nor mutation detection is error-free, and mutations were sought only in the *MLH1* and *MSH2* genes. Two founder mutations account for over half of all hereditary nonpolyposis colorectal cancer mutations in this population.[28] These mutations can be easily screened for in large numbers of samples.

cancer.[7] Familial adenomatous polyposis and most sporadic cases may be considered a paradigm for the first, or distal, class of colorectal cancers, whereas hereditary nonpolyposis colorectal cancer more clearly represents the second, or proximal, class.[7]

A hallmark of tumors in hereditary nonpolyposis colorectal cancer is microsatellite instability.[19–21] Microsatellites are genomic regions in which short DNA sequences or a single nucleotide is repeated. There are hundreds of thousands of microsatellites in the human genome. During DNA replication, mutations occur in some microsatellites owing to the misalignment of their repetitive subunits and result in contraction or elongation ("instability"). These abnormalities are usually repaired by the mismatch-repair proteins. However, repair

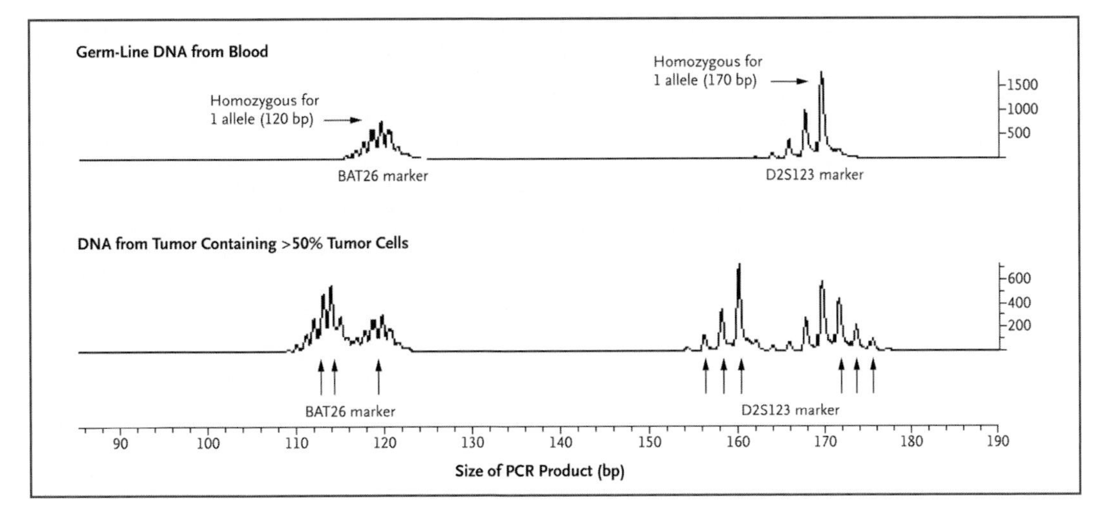

Figure 8.2 Detection of Microsatellite Instability with the Use of Fluorescent Labeling of Polymerase-Chain-Reaction (PCR) Products Analyzed in an Automatic Sequencer. Two markers are analyzed in the same track: the mononucleotide repeat marker BAT26 is shown on the left, and the dinucleotide marker D2S123 is shown on the right. The upper tracing is from germ-line DNA from blood. The lower tracing is from DNA extracted from a histologic section of a tumor containing more than 50 percent tumor cells. For marker BAT26, germ-line DNA shows a single peak, indicating that the patient is homozygous for this marker (arrow). Tumor DNA shows, in addition to the normal allele (single arrow), a new allele (double arrows) that has lost approximately five nucleotides. This constitutes microsatellite instability. For marker D2S123, germ-line DNA is homozygous, whereas tumor DNA shows two new alleles (triple arrows), one with a loss of approximately 10 nucleotides (left) and one with a gain of 2 nucleotides (right). Thus, the tumor shows microsatellite instability with both markers. All peaks display "stutter"—that is, small amounts of material with a gain or a loss of one or a few nucleotides. This is a normal phenomenon.

is inefficient in tumors with a deficiency of these proteins. Typically, in such tumor cells, half or more of all microsatellites have mutations (contraction or elongation), so microsatellite instability serves as an excellent, easy-to-evaluate marker of mismatch-repair deficiency (Fig. 8.1). Since microsatellite instability is found in virtually all hereditary nonpolyposis colorectal cancers,[29] we consider it unnecessary to search for germ-line mutations in mismatch-repair genes (e.g., *MSH2* and *MLH1*) in patients whose tumors do not have microsatellite instability. An exception is found in families with the *MSH6* mutation, in which microsatellite instability may or may not be present.[30,31] Most microsatellites occur in noncoding DNA; therefore, contractions or elongations are believed to have little or no effect on protein function. However, there are genes that have microsatellites in their coding regions (Fig. 8.2), and microsatellite instability will thus lead to altered proteins.

FAMILIAL ADENOMATOUS POLYPOSIS

Clinical and Molecular Features

Multiple colonic adenomas occur at an early age in patients with familial adenomatous polyposis, occasionally during the preteen years, and proliferate throughout the colon, with malignant degeneration in most patients by the age of 40 to 50 years. Patients who have an *APC* mutation or who have one or more first-degree relatives with familial adenomatous polyposis or an identified *APC* mutation (or both) are at high risk and should be screened with flexible sigmoidoscopy by the age of 10 to 12 years. Patients with colonic polyps, a verified *APC* germ-line mutation, or both will require annual endoscopic examination. However, as the disease advances, as is often the case in the late teens and early 20s, too many colonic polyps may be present for adequate and safe colonoscopic polypectomy; when this occurs, prophylactic subtotal colectomy followed by annual endoscopy of the remaining rectum is recommended.

Upper endoscopy is also necessary because of the potential for adenomas, which increase the risk of cancer of the stomach. Although cancers of the stomach are uncommon in whites, they are of particular concern to families with familial adenomatous polyposis in Korea and Japan.[32] Adenomas in the duodenum, which carry a risk of a periampullary carcinoma, and in the remainder of the small intestine are more common.[33] There is limited knowledge about the causation, prevention, and management of duodenal polyposis in familial adenomatous polyposis. However, there is a strong association with stage IV periampullary adenomas, which pose a high lifetime risk of periampullary carcinoma in patients with familial adenomatous polyposis.[34] Even though the efficacy of screening is yet to be fully demonstrated, Burke[33] recommends upper endoscopic screening with forward- and side-viewing endoscopes for all those with a family history of familial adenomatous polyposis.

Desmoids also appear frequently in patients with familial adenomatous polyposis and are often induced by surgery.[35,36] Ideally, prophylactic colectomy should be delayed unless there are too many colonic adenomas to manage safely. Elective surgical procedures should be avoided whenever this is possible. Other, less common tumors that may occur in families with familial adenomatous polyposis include papillary thyroid carcinoma, sarcomas, hepatoblastomas, pancreatic carcinomas, and medulloblastomas of the cerebellar–pontine angle of the brain.[36–41] With the exception of papillary thyroid carcinoma, screening for these tumors is difficult and therefore not generally performed.

The penetrance of germ-line mutations that increase the risk of colorectal cancer varies.[38,40,42] It is 10 to 20 percent for the I1307K *APC* polymorphism, which occurs predominantly in Ashkenazi Jews (Fig. 8.3). In contrast, penetrance approaches 100 percent in classic familial adenomatous polyposis,[47] caused by truncating germ-line mutations of the *APC* gene.

Genetic Testing

Genetic counseling should be performed by a genetic counselor or medical geneticist before DNA is collected and at the time of the disclosure of test results. We recommend discussing the matter in depth with the parents of patients who are younger than 18 years, as well as with the patients themselves, since polyps may occur in the preteen and teen years, and cancer may occur relatively early in some of these patients. It is important for the counselor to know whether the *APC* mutation is present, and if so, its probable penetrance, particularly in patients with attenuated familial adenomatous polyposis.[10,37]

Chemoprevention

Patients with familial adenomatous polyposis who were treated with 400 mg of celecoxib, a selective inhibitor of cyclooxygenase-2, twice a day for six months had a 28.0 percent

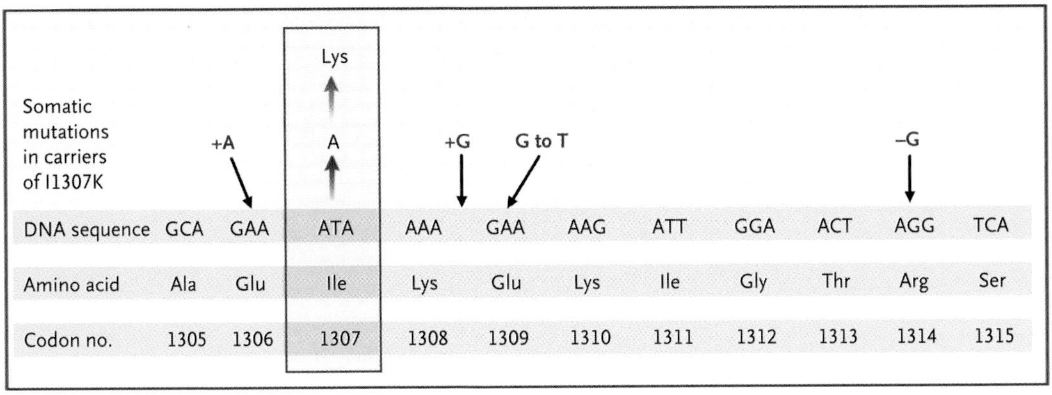

Somatic mutations in carriers of I1307K		+A	Lys A	+G	G to T					−G	
DNA sequence	GCA	GAA	ATA	AAA	GAA	AAG	ATT	GGA	ACT	AGG	TCA
Amino acid	Ala	Glu	Ile	Lys	Glu	Lys	Ile	Gly	Thr	Arg	Ser
Codon no.	1305	1306	1307	1308	1309	1310	1311	1312	1313	1314	1315

Figure 8.3 The I1307K Germ-Line Mutation (Polymorphism) of the Adenomatous Polyposis Coli (*APC*) Gene. Shown here is the DNA sequence of codons 1305 through 1315 of the *APC* gene. Below each codon is the encoded amino acid and the number of the codon. The germ-line mutation of codon 1307 shown in the blue box is a change from T to A that changes an ATA encoding isoleucine (abbreviated I in the one-letter system) to an AAA encoding lysine (abbreviated K). Thus, the designation for the mutation is I1307K. This change is believed to be a neutral variant — that is, it does not alter the function of the APC protein; hence, it may be called both a mutation and a polymorphism.[43] Approximately 6 percent of Ashkenazi Jews and a smaller proportion of other Jews are carriers of the I1307K mutation or polymorphism; it has not been seen in non-Jews.[43–45] As compared with non-carriers, carriers have approximately twice the risk of colorectal cancer.[43] The T-to-A change results in a stretch of eight adenosines (AAAAAAAA) that is believed to increase the risk of somatic mutations as a result of slippage during replication. Examples of these somatic changes in colonic tumors are shown in red above the sequence. For instance, an addition of one A (+A) has been seen in the affected allele of many carriers. The addition or loss of a nucleotide causes a frame shift and loss of function of APC, constituting an important somatic event in tumor initiation.[43,46]

reduction in the mean number of colorectal polyps ($P = 0.003$), as compared with patients in the placebo group.[48] However, polyps may return while the patient is taking nonsteroidal antiinflammatory drugs. In one study, regression of colonic adenomas occurred in all patients after six months of sulindac (200 mg per day) ($P < 0.02$).[49] However, after a mean of 48.6 months, the number and size of the polyps increased. At a dose of 200 mg, sulindac did not influence the progression of polyps toward a malignant pattern.[49] There is hope that large, ongoing chemoprevention trials will provide concrete clues as to the future of antiinflammatory agents in the prevention of polyps and cancer.[50,51] Currently, none of these chemoprevention strategies should replace screening, although they may delay prophylactic colectomy.[52]

HEREDITARY NONPOLYPOSIS COLORECTAL CANCER

Clinical Features

Hereditary nonpolyposis colorectal cancer, also referred to as the Lynch syndrome, is the most common form of hereditary colorectal cancer. Multiple generations are affected with colorectal cancer at an early age (mean, approximately 45 years) with a predominance of right-sided colorectal cancer (approximately 70 percent proximal to the splenic flexure). There is an excess of synchronous colorectal cancer (multiple colorectal cancers at or within six months after surgical resection for colorectal cancer) and metachronous colorectal cancer (colorectal cancer occurring more than six months after surgery).[7] In addition, there is an

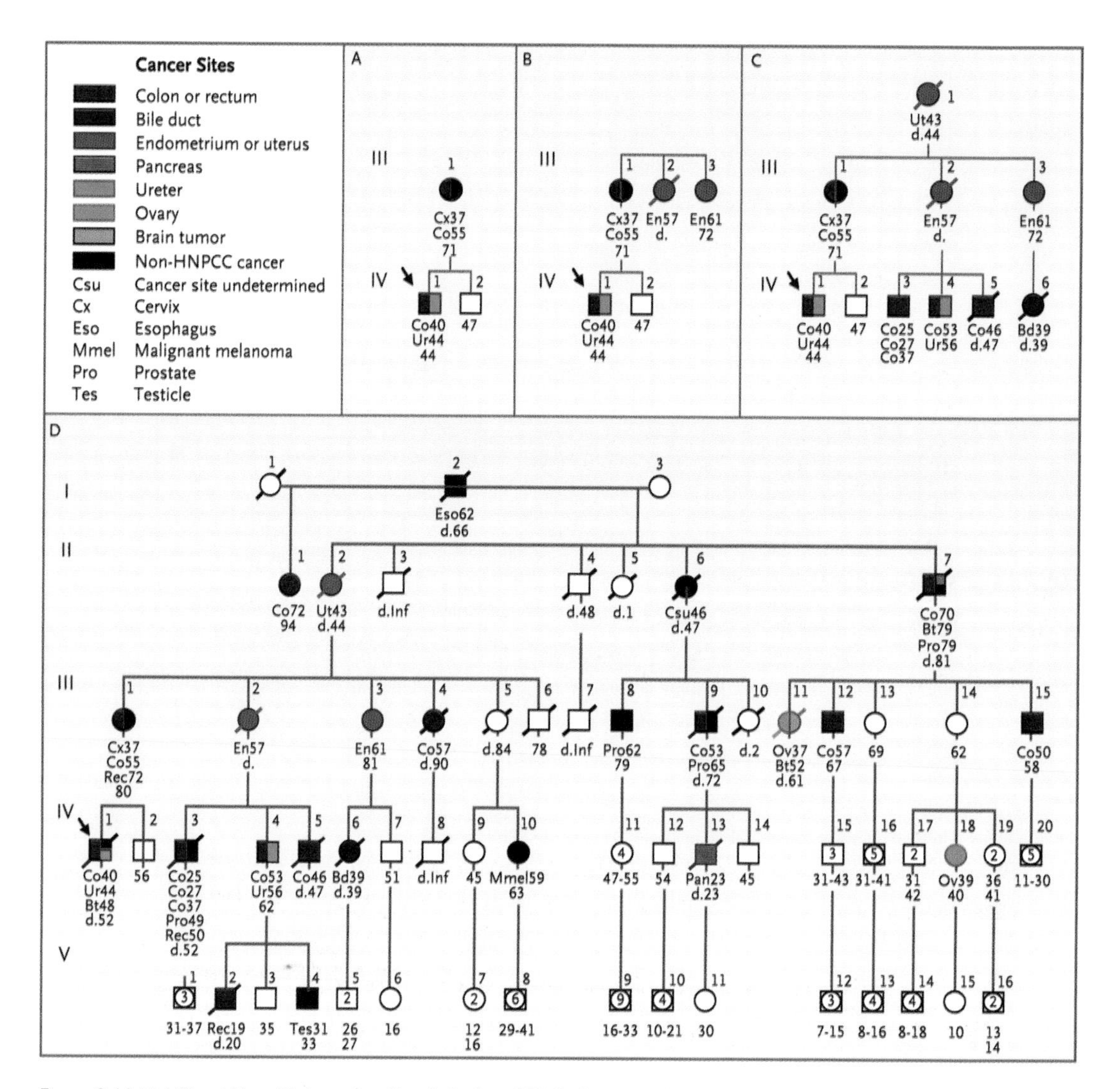

Figure 8.4 Initial (Panel A) and Subsequent (Panels B, C, and D) Evaluations of a Pedigree with Classic Hereditary Nonpolyposis Colorectal Cancer. Panel A shows the initial assessment of what turned out to be a family with classic hereditary nonpolyposis colorectal cancer (HNPCC). The proband (Subject IV-1; arrow) had early-onset (40 years) colorectal carcinoma and carcinoma of the ureter. These findings by themselves are highly significant clinically. However, his mother (Subject III-1) had uterine cervical carcinoma at the age of 37 years, a tumor not associated with the syndrome, but had colorectal carcinoma at the age of 55 years. In Panel B, further inquiry indicated that the proband's mother (Subject III-1) had two sisters with endometrial carcinoma at the ages of 57 (Subject III-2) and 61 (Subject III-3). This pattern, notwithstanding the Amsterdam criteria, would be sufficient for a diagnosis of hereditary nonpolyposis colorectal cancer. In Panel C, extending the pedigree further showed that one of these maternal aunts (Subject III-2) had three sons with cancer, one with early-onset metachronous colon cancer (Subject IV-3), a second with colon cancer and carcinoma of the ureter (Subject IV-4), and the third with colon cancer alone (Subject IV-5). The other aunt (Subject III-3) had a daughter (Subject IV-6) with cancer of the bile duct. These findings provide strong evidence in support of the diagnosis of hereditary nonpolyposis colorectal cancer. In Panel D, the full pedigree shows findings that continue to support a diagnosis of hereditary nonpolyposis colorectal cancer.[55]

Squares denote male family members; circles female family members; symbols with a slash deceased family members, with the age at death (d) given below each symbol; open symbols unaffected family members; bicolored symbols family members with multiple primary cancers; squares containing numbers the number of unaffected male progeny; circles containing numbers the number of unaffected female progeny; and combined symbols containing numbers the number of unaffected progeny of both sexes. The types of primary cancer and the age (in years) at diagnosis are listed below the symbols, and the bottom-most number is the current age or the age at death. Inf denotes in infancy.

excess of extracolonic cancers—namely, carcinoma of the endometrium (second only to colorectal cancer in frequency), ovary, stomach (particularly in Asian countries such as Japan and Korea[32]), small bowel, pancreas, hepatobiliary tract, brain, and upper uroepithelial tract.[12,53] There is also an apparent statistically significant decrease in the risk of lung cancer,[12] which, while not proved, merits further research. Patients with hereditary nonpolyposis colorectal cancer may also have sebaceous adenomas, sebaceous carcinomas, and multiple keratoacanthomas, findings consonant with Torre's syndrome variant.[7,54]

Figure 8.4 depicts the evaluation of a family with hereditary nonpolyposis colorectal cancer from initial ascertainment to completion. The figure illustrates the advantage of seeking a more extensive family history when initial information is limited but includes clinical findings suggestive of hereditary nonpolyposis colorectal cancer. For example, two siblings may have colorectal cancer of the proximal colon before the age of 30 years in the absence of multiple colonic adenomas. However, their parents may have died at an early age of causes other than cancer and other relatives with potentially valuable genetic information may simply not be available for testing. Although neither of these clinical scenarios fulfills the Amsterdam I or II criteria for hereditary nonpolyposis colorectal cancer (Table 8.1), the clinician may prudently wish to err on the side of caution. Additional study of the tumor should include microsatellite-instability testing in at least one of the colorectal cancers or a search for a mutation in a mismatch-repair gene, such as *MSH2* or *MLH1*, in the resected tumor.

Pathological Features

As compared with sporadic colorectal cancer, tumors in hereditary nonpolyposis colorectal cancer are more often poorly differentiated, with an excess of mucoid and signet-cell

Table 8.1 Amsterdam I and Amsterdam II Criteria*

Amsterdam I criteria

At least three relatives must have histologically verified colorectal cancer:
- One must be a first-degree relative of the other two.
- At least two successive generations must be affected.
- At least one of the relatives with colorectal cancer must have received the diagnosis before the age of 50 years.
- Familial adenomatous polyposis must have been excluded.

Amsterdam II criteria

At least three relatives must have a cancer associated with hereditary non-polyposis colorectal cancer (colorectal, endometrial, stomach, ovary, ureter or renal-pelvis, brain, small-bowel, hepatobiliary tract, or skin [sebaceous tumors]):
- One must be a first-degree relative of the other two.
- At least two successive generations must be affected.
- At least one of the relatives with cancer associated with hereditary nonpolyposis colorectal cancer should have received the diagnosis before the age of 50 years.
- Familial adenomatous polyposis should have been excluded in any relative with colorectal cancer.
- Tumors should be verified whenever possible.

*The Amsterdam I and II criteria are from Vasen et al.[56,57]

features, a Crohn's-like reaction (lymphoid nodules, including germinal centers, located at the periphery of infiltrating colorectal carcinomas), and the presence of infiltrating lymphocytes within the tumor.[58–61]

Accelerated Carcinogenesis

Accelerated carcinogenesis occurs in hereditary nonpolyposis colorectal cancer. In this setting, a tiny colonic adenoma may emerge as a carcinoma within 2 to 3 years, as opposed to the 8 to 10 years this process may take in the general population.[7,61] This rapid growth leads us to recommend annual colonoscopy, as discussed below.

Features of Pedigrees

The original definitions based on clinical and pedigree criteria such as the more stringent Amsterdam I criteria[56] or the less stringent Amsterdam II criteria[57] remain valid (Table 8.1). However, in many situations, even if the criteria are not met, the occurrence of cancers associated with hereditary nonpolyposis colorectal cancer, especially in small families, should alert the clinician to the possibility of hereditary nonpolyposis colorectal cancer, as should cancer at a very early age or multiple cancers in one person (Fig. 8.5).

Incidence and Molecular Screening

When the Amsterdam criteria (Table 8.1) are used to determine what proportion of all colorectal cancers are due to hereditary nonpolyposis colorectal cancer, estimates range from 1 to 6 percent.[7,22,23,62] Molecular screening of all patients with colorectal cancer for hereditary nonpolyposis colorectal cancer is now both feasible and desirable. Such screening has suggested that upward of 3 percent of all such patients have hereditary nonpolyposis colorectal cancer (Fig. 8.1). In one study, the mean age at presentation with hereditary nonpolyposis colorectal cancer diagnosed by molecular screening was 54 years old; the study included several patients over 60 years of age, and some had a minimal family history of cancer.[22,23] If further studies confirm these findings, the age at onset may prove older than the mean of 45 years in cases ascertained on the basis of family-history criteria. For this reason, we recommend that whenever population-based screening is performed, it include all patients with colorectal cancer irrespective of age and family history.

Analysis of mutations in mismatch-repair genes has provided estimates of the proportion of such mutations in families with a history consonant with hereditary nonpolyposis colorectal cancer. These estimates range from 40 to 80 percent for families meeting the Amsterdam I criteria and from 5 to 50 percent for families meeting the Amsterdam II criteria.[62,63] Among such families, as well as in other families whose history is consistent with the presence of hereditary nonpolyposis colorectal cancer but who do not meet these formal criteria, some families will not harbor a known mismatch-repair mutation. This is consistent with the notions that in such families other, as yet undiscovered genes may be responsible for the syndrome and that the aggregation of cancers may be caused by environmental factors or be due to chance.[5]

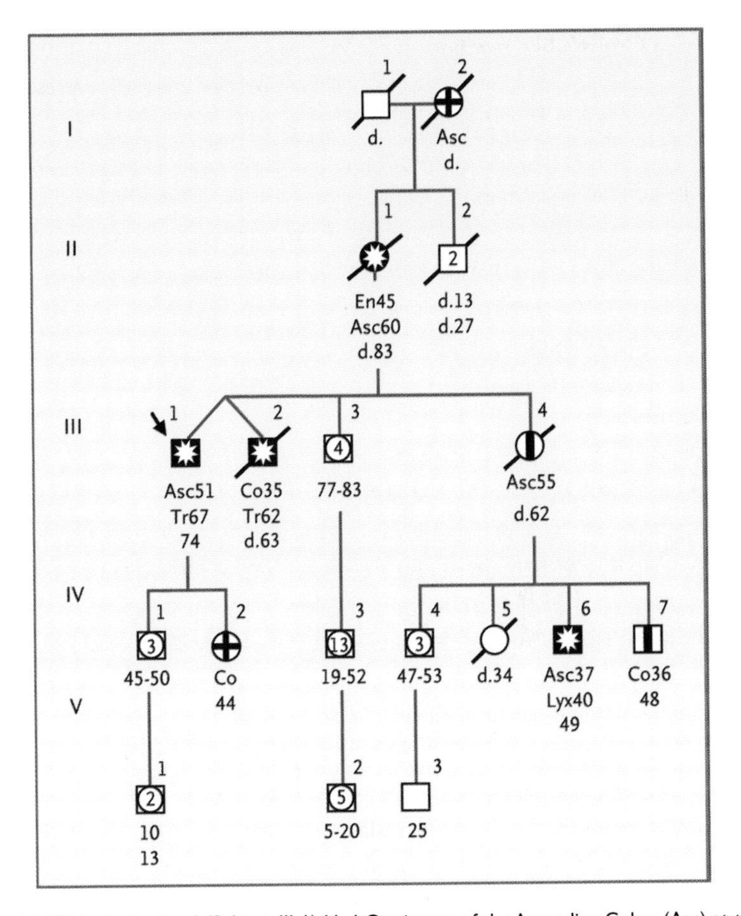

Figure 8.5 Pedigree in Which the Proband (Subject III-1) Had Carcinoma of the Ascending Colon (Asc) at the Age of 51 Years and a Second Primary Carcinoma of the Transverse Colon (Tr) at the Age of 67 Years. The proband's fraternal twin brother (Subject III-2) had colon cancer (Co)—precise site unknown—at the age of 35 years followed by a second primary cancer of the transverse colon at the age of 62 years. Their sister (Subject III-4) had cancer of the ascending colon at the age of 55 years. Their mother (Subject II-1) had carcinoma of the endometrium (En) at the age of 45 years and carcinoma of the ascending colon at the age of 60 years, and the proband's maternal grandmother had carcinoma of the ascending colon. The proband's daughter had colon cancer at the age of 44 years, and a nephew had carcinoma of the ascending colon at the age of 37 years and carcinoma of the larynx (Lyx) at the age of 40 years. The progeny in the direct genetic lineage of Subjects III-1 and III-4 merit intensive surveillance and would be candidates for genetic testing. The mutation discovered in the family is a missense mutation involving *MLH1*. Squares denote male family members; circles female family members; symbols with a slash deceased family members, with the age at death (d) given below each symbol; open symbols unaffected family members; solid symbols with a star family members with pathological evidence of multiple primary cancers, with the age at diagnosis shown to the right of the types of cancer; a divided symbol a family member with cancer established on the basis of the family history; a symbol with a cross a family member whose cause of death was determined by examining the death certificate or medical records; and combined symbols containing numbers the number of unaffected progeny of both sexes. Bottom-most numbers are current ages.

Genes and Germ-Line Mutations

Hereditary nonpolyposis colorectal cancer is caused by a germ-line mutation in any of the mismatch-repair genes listed in Table 8.2. As of this writing, two genes, *MLH1* and *MSH2*, account for almost 90 percent of all identified mutations. *MSH6* accounts for almost 10 percent, but its share of typical as opposed to less typical hereditary

Table 8.2 Number of Different Germ-Line Mutations and Polymorphisms Identified in Patients with Hereditary Nonpolyposis Colorectal Cancer*			
Gene	Total No. of Mutations†	No. of Missense Mutations (% of total)	No. of Polymorphisms‡
MLH1	164	47 (29)	20
MSH2	121	19 (16)	24
MSH6	31	12 (39)	43
PMS1	1	0	0
PMS2	5	1 (20)	5
Total	322	79 (25)	92

*The mutations are from the data maintained by the International Collaborative Group on Hereditary Non-Polyposis Colorectal Cancer (http://www.nfdht.nl). The data base also lists 10 mutations in the *MLH3* gene; all but 1 are missense mutations and have so far been reported by a single laboratory.[64] Their putative pathogenetic role remains to be determined.

†The mutations listed are considered disease-causing.

‡The polymorphisms listed are not considered disease-causing.

nonpolyposis colorectal cancer remains to be determined.[30,31] It is usually sufficient first to screen patients for *MLH1* and *MSH2* and then to test other genes only if mutations are not found in these two.

Assessing the Pathogenicity of Mutations

All genomic coding changes are potentially deleterious. However, as opposed to nonsense mutations (which create a stop codon or lead to a frame shift) or those that cause abnormal splicing, missense mutations (which lead to the substitution of an amino acid) are usually not considered a priori pathogenic. Of all mutations identified in *MLH1* and *MSH2*, 29 percent and 16 percent, respectively, are missense mutations. Missense mutations make the interpretation of genotypic data difficult. The mutation data base maintained by the International Collaborative Group on Hereditary Non-Polyposis Colorectal Cancer is an important primary reference (http://www.nfdht.nl). Immunohistochemical analysis of mismatch-repair proteins in the tumor can provide clues as to which mismatch-repair gene is involved in tumor pathogenesis if staining for one of the proteins is weak or absent.[25,65]

Sources of Underdiagnosis

Previous estimates of the frequency of hereditary nonpolyposis colorectal cancer were most likely low. Most analyses of mutations to date have not included analysis of *MSH6*, which undoubtedly causes hereditary nonpolyposis colorectal cancer or a predisposition to an atypical and more benign form of this syndrome.[30] Moreover, conventional mutation analysis overlooks some mutations that can be detected only when the two alleles are studied separately, with the use of more sophisticated techniques.[66] Such techniques permit the detection of several types of mutation that elude conventional mutation analysis, mainly mutations in control regions or introns that affect transcription or splicing.[67] Finally, large deletions in the *MSH2* gene are more common than previously thought and can be detected by Southern hybridization.[68]

Surveillance for Cancer

In patients with hereditary nonpolyposis colorectal cancer, annual full colonoscopy, initiated between the ages of 20 and 25 years, is recommended for those with strong clinical evidence or documented germ-line mutations in *MLH1*, *MSH2*, or *MSH6* (or a combination). Although less frequent colonoscopy (every three years) has been suggested in a consensus statement,[69] we believe this would lead to missed colorectal cancers, given the phenomenon of accelerated carcinogenesis in such cancers.[7,60,61] Extracolonic screening, particularly of the endometrium and ovary, the sites of the second and third most common cancers in this disorder, is indicated in patients with hereditary nonpolyposis colorectal cancer. With respect to the endometrium, annual transvaginal ultrasonography and endometrial aspiration for pathological assessment should be begun at the age of 30 years and repeated annually. In the case of the ovary, this evaluation should include transvaginal ovarian ultrasonography and CA-125 screening, also beginning at the age of 30 years. Patients should be aware of the low sensitivity and specificity of surveillance methods for ovarian cancer. Screening at other sites, such as the upper uroepithelial tract and stomach (particularly in natives of Korea[32] or Japan or in a family with an excess number of cancers at these extracolonic sites) must be considered, but it is difficult.

Efficacy of Surveillance

The efficacy of surveillance for colorectal cancer in families with hereditary nonpolyposis colorectal cancer was evaluated in a controlled clinical trial extending over a 15-year period.[70] The study concluded that screening for colorectal cancer at three-year intervals more than halves the risk of colorectal cancer, prevents deaths from colorectal cancer, and decreases the overall mortality rate by about 65 percent in such families. The relatively high incidence of colorectal cancer (albeit nonfatal cases) even among these frequently screened subjects is an argument for shorter screening intervals, such as one year. Prophylactic subtotal colectomy, prophylactic total abdominal hysterectomy, and bilateral salpingo-oophorectomy are presented as options to selected patients.[7,71]

The identification of hereditary nonpolyposis colorectal cancer can be lifesaving, since it can lead to the early detection of cancer.[70,72] This effect was quantified in a study by Ramsey et al.,[73] a cost-effectiveness analysis comparing standard care with a process that included the application of the Bethesda guidelines (which identify the colorectal tumors to test for microsatellite instability),[74] followed by testing of the tumor for microsatellite instability, germ-line testing, and lifelong screening for colorectal cancer among carriers of mutations. The cost of screening was $7,556 per year of life gained when patients with cancer and their siblings and children were considered together.[73]

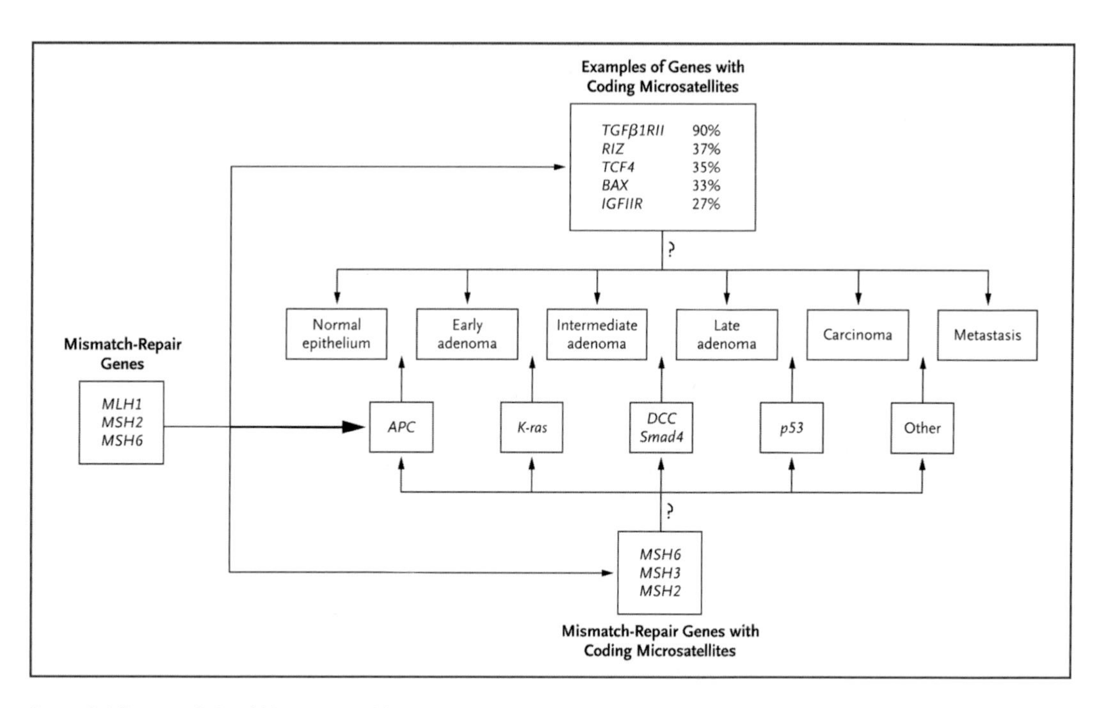

Figure 8.6 Putative Role of Mutations in Mismatch-Repair Genes. A mutation in one of the three main mismatch-repair genes leads to a deficiency of mismatch-repair proteins, which promotes mutations in some of the traditional genes, such as the adenomatous polyposis coli gene *(APC)* and *K-ras*. Genes with coding microsatellites accumulate frame-shift mutations and lose function, further affecting, and perhaps speeding up, the evolution of cancer. They may also affect the organ specificity of hereditary nonpolyposis colorectal cancers. The approximate percentages of all colorectal-cancer tumors with mismatch-repair deficiency that harbor these frame-shift mutations are shown in the box at the top.[56,76,79,80] Three mismatch-repair genes that have coding microsatellites are shown at the bottom.[81,82] Somatic frame-shift mutations occur in these genes, but their role remains unclear. *TGFβ1RII* denotes transforming growth factor β1 receptor II, *RIZ* retinoblastoma-protein–interacting zinc finger, *TCF4* transcription factor 4, *BAX* BCL-2–associated X, *IGFIIR* insulin-like growth factor II receptor, and *DCC* deleted in colorectal cancer.

Somatic Mutations and the Progression to Cancer

The multigene, clonal evolution, and selection model of the initiation and progression of cancer proposed by Fearon and Vogelstein originally identified *APC*, genes on 18q, *Ras*, and *p53 (TP53)* as the genes in which mutations or epigenetic dysregulation contributes to the evolution of colon cancer.[75] Although later studies have confirmed this model, many additional genes are also involved.[46,76] What is the role of a mismatch-repair deficiency in this model? In colorectal tumors with a deficiency of mismatch-repair protein, all the named components are involved,[19] but probably to different degrees.[77] It appears that the genetic pathways are the same even though the involvement of the different genes varies.[78] Figure 8.6 shows the putative role of mutations in mismatch-repair genes.

Role of Epigenetics

Methylation of the CpG sites in the promoter region of *MLH1* silences its transcription and, when both alleles are affected, leads to a typical mismatch-repair deficiency.[83,84] This epigenetic change is not heritable and accounts for the majority of all sporadic colorectal cancers that are positive for microsatellite instability.[85–87] These tumors typically affect patients older than 60 years of age and women, are right-sided, and carry the same histologic and prognostic hallmarks as hereditary nonpolyposis colorectal cancers.[15]

HAMARTOMATOUS POLYPOSIS SYNDROMES

The differential diagnosis of juvenile polyposis syndrome includes Cowden's disease, the Bannayan–Ruvalcaba–Riley syndrome, and the Peutz–Jeghers syndrome, and there are often only very subtle clinical distinctions among them. Hence, the emerging evidence of their molecular bases may allow more precise distinctions to be made among these syndromes (Table 8.3). For example, germ-line mutations in *PTEN*, a protein tyrosine phosphatase gene, have been identified in Cowden's disease and Bannayan–Ruvalcaba–Riley syndrome that show that the two syndromes may be allelic and "might even be one and the same syndrome along a broad spectrum." Eng and Ji discuss the problem of phenotypic features that may be shared by the various hamartomatous syndromes, thereby contributing to the complexity of clinical diagnosis.[94] They suggest referring such patients to physicians with extensive experience with these disorders.

Table 8.3 Hamartomatous Polyposis Syndromes

Syndrome	Phenotype	Mutant Gene*
Peutz–Jeghers syndrome	Perioral pigmentations, pigmentations of fingers, upper and lower gastrointestinal hamartomatous lesions, small-bowel and pancreatic cancer, colorectal cancer and sex-cord tumors with annular tubules of the ovary	*LKB1 (STK11)*[88]
Familial juvenile polyposis	Gastrointestinal hamartomatous polyps, increased risk of gastrointestinal cancer (stomach, colorectum)†	*Smad4 (DPC4)*[89] *BMPRIA*[90] *PTEN*[91]
Cowden's disease‡	Colonic hamartomatous polyps, benign and malignant neoplasms of the thyroid, breast, uterus, and skin (multiple trichilemmomas)§	*PTEN (MMAC1, DEP1)*[92,93]
Bannayan–Ruvalcaba–Riley syndrome	Microcephaly, fibromatosis, hamartomatous polyposis, hemangiomas, speckled penis§	*PTEN*

*Alternative names are given in parentheses.

†The diagnosis is made only when features pathognomonic of the other syndromes are not present.

‡Finding a germ-line mutation in *PTEN* provides molecular evidence of Cowden's disease, but the absence of an identifiable *PTEN* mutation is non-diagnostic.

§A firm association with colorectal cancer has yet to be identified.

Prospects for Prevention and Treatment

Morbidity and mortality from hereditary forms of colorectal cancer should be reduced once a patient's familial or hereditary risk is established and a highly targeted program of cancer surveillance and management is undertaken.[69,70,73] Prevention will be aided by the identification of the causative germ-line mutation in a patient's family, thus confirming the risk. Cancer prevention, particularly among patients with familial adenomatous polyposis and hereditary nonpolyposis colorectal cancer, will be most effective when physicians understand the natural history and the molecular bases of these disorders. They must recognize the need for genetic counseling before DNA testing is performed and at the time the test results are disclosed. A vexing problem is the perception of many high-risk patients that participating in genetic-testing, clinical, and research programs, which can contribute to the identification and ultimate prevention and reduction in morbidity and mortality from hereditary cancer syndromes, will result in discrimination by insurance companies or employers.[9,95–97] Legislative bodies need to enact laws that will protect such patients from potential discrimination.[95–98]

In addition to diagnostic methods, physicians must also be familiar with the available screening methods and with the options for surgical prophylaxis, particularly prophylactic colectomy in patients with familial adenomatous polyposis and prophylactic colectomy and prophylactic bilateral salpingo-oophorectomy (the latter when childbearing is complete) in patients with hereditary nonpolyposis colorectal cancer. Technologic advances in both cancer screening and the identification of biologic markers of cancer susceptibility, such as microsatellite instability, and ultimately specific germ-line testing, will expedite attempts to achieve these cancer-prevention goals. Pharmacologic treatment that is based on molecular-targeting strategies holds great promise.[99,100]

Finally, molecular genetic research on hereditary forms of colorectal cancer must continue to search for new mutations in these heterogeneous disorders. For example, researchers have described germ-line mutations in *MYH* in classic familial adenomatous polyposis coli showing apparent autosomal recessive inheritance.[101,102]

Acknowledgments

Supported by revenue from Nebraska cigarette taxes awarded to Creighton University by the Nebraska Department of Health and Human Services and by grants (1U01 CA86389-01, P30 CA16058, and R01 CA67941) from the National Institutes of Health and by Folkhälsan Institute of Genetics, Helsinki, Finland.

The contents of this chapter are solely the responsibility of the authors and do not necessarily represent the official views of the State of Nebraska or the Nebraska Department of Health and Human Services.

We are indebted to Trudy Shaw, M.A., for her faithful and diligent technical assistance throughout the development of this manuscript, and to Jane Lynch, B.S.N., who was a constant source of help in our family studies and in portions of the writing of this article.

References

1. Jemal A, Thomas A, Murray T, Thun M. Cancer statistics, 2002. CA Cancer J Clin 2002; 52:23–47. [Errata, CA Cancer J Clin 2002; 52:119, 181–2.]

2. Polednak AP. Do physicians discuss genetic testing with family-history-positive breast cancer patients? Conn Med 1998; 62:3–7.

3. Cho MK, Sankar P, Wolpe PR, Godmilow L. Commercialization of *BRCA1/2* testing: practitioner awareness and use of a new genetic test. Am J Med Genet 1999; 83:157–63.

4. Giardiello FM, Brensinger JD, Petersen GM, et al. The use and interpretation of commercial *APC* gene testing for familial adenomatous polyposis. N Engl J Med 1997; 336:823–7.

5. Vogelstein B, Kinzler KW, eds. The genetic basis of human cancer. New York: McGraw-Hill, 1998.

6. Eng C, Hampel H, de la Chapelle A. Genetic testing for cancer predisposition. Annu Rev Med 2001; 52:371–400. [Erratum, Annu Rev Med 2002; 53:xi.]

7. Lynch HT, de la Chapelle A. Genetic susceptibility to non-polyposis colorectal cancer. J Med Genet 1999; 36:801–18.

8. Lynch HT. Family information service and hereditary cancer. Cancer 2001; 91:625–8.

9. Aktan-Collan K, Mecklin JP, Jarvinen HJ, et al. Predictive genetic testing for hereditary non-polyposis colorectal cancer: uptake and long-term satisfaction. Int J Cancer 2000; 89:44–50.

10. Lynch HT, Smyrk T, McGinn T, et al. Attenuated familial adenomatous polyposis (AFAP): a phenotypically and genotypically distinctive variant of FAP. Cancer 1995; 76:2427–33.

11. Lynch HT, Lynch JF, Casey MJ, Bewtra C, Narod SA. Genetics of gynecological cancer. In: Hoskins WJ, Perez CA, Young RC, eds. Principles and practice of gynecologic oncology. 3rd ed. Philadelphia: Lippincott Williams & Wilkins, 2000:29–53.

12. Watson P, Lynch HT. The tumor spectrum in HNPCC. Anticancer Res 1994; 14:1635–9.

13.Watson P, Lin KM, Rodriguez-Bigas MA, et al. Colorectal carcinoma survival among hereditary non-polyposis colorectal carcinoma family members. Cancer 1998; 83:259–66.

14. Sankila R, Aaltonen LA, Jarvinen HJ, Mecklin J-P. Better survival rates in patients with MLH1-associated hereditary colorectal cancer. Gastroenterology 1996; 110:682–7.

15. Gryfe R, Kim H, Hsieh ETK, et al. Tumor microsatellite instability and clinical outcome in young patients with colorectal cancer. N Engl J Med 2000; 342:69–77.

16. Branch P, Bicknell DC, Rowan A, Bodmer WF, Karran P. Immune surveillance in colorectal carcinoma. Nat Genet 1995; 9:231–2.

17. Shashidharan M, Smyrk T, Lin KM, et al. Histologic comparison of hereditary nonpolyposis colorectal cancer associated with MSH2 and MLH1 and colorectal cancer from the general population. Dis Colon Rectum 1999; 42:722–6.

18. Boman BM, Fry RD, Curran W, et al. Unique immunohistochemical features of MSI-classified tumors from HNPCC patients. Prog Proc Am Soc Clin Oncol 1999; 18:236a. abstract.

19. Aaltonen LA, Peltomaki P, Leach FS, et al. Clues to the pathogenesis of familial colorectal cancer. Science 1993; 260:812–6.

20. Peltomaki P, Lothe RA, Aaltonen LA, et al. Microsatellite instability is associated with tumors that characterize the hereditary non-polyposis colorectal carcinoma syndrome. Cancer Res 1993; 53:5853–5.

21. Boland CR, Thibodeau SN, Hamilton SR, et al. A National Cancer Institute workshop on microsatellite instability for cancer detection and familial predisposition: development of international criteria for the determination of microsatellite instability in colorectal cancer. Cancer Res 1998; 58:5248–57.

22. Aaltonen LA, Salovaara R, Kristo P, et al. Incidence of hereditary nonpolyposis colorectal cancer and the feasibility of molecular screening for the disease. N Engl J Med 1998; 338:1481–7.

23. Salovaara R, Loukola A, Kristo P, et al. Population-based molecular detection of hereditary nonpolyposis colorectal cancer. J Clin Oncol 2000; 18:2193–200. [Erratum, J Clin Oncol 2000; 18:3456.]

24. Lothe RA, Peltomaki P, Meling GI, et al. Genomic instability in colorectal cancer: relationship to clinicopathological variables and family history. Cancer Res 1993; 53:5849–52.

25. Thibodeau SN, French AJ, Cunningham JM, et al. Microsatellite instability in colorectal cancer: different mutator phenotypes and the principal involvement of hMLH1. Cancer Res 1998; 58:1713–8.

26. Zhou X-P, Hoang J-M, Li Y-J, et al. Determination of the replication error phenotype in human tumors without the requirement for matching normal DNA by analysis of mononucleotide repeat microsatellites. Genes Chromosomes Cancer 1998; 21:101–7.

27. de la Chapelle A. Testing tumors for microsatellite instability. Eur J Hum Genet 1999; 7:407–8.

28. Nyström-Lahti M, Kristo P, Nicolaides NC, et al. Founding mutations and Alu-mediated recombination in hereditary colon cancer. Nat Med 1995; 1:1203–6.

29. Aaltonen LA, Peltomaki P, Mecklin JP, et al. Replication errors in benign and malignant tumors from hereditary nonpolyposis colorectal cancer patients. Cancer Res 1994; 54:1645–8.

30. Miyaki M, Konishi M, Tanaka K, et al. Germline mutation of *MSH6* as the cause of hereditary nonpolyposis colorectal cancer. Nat Genet 1997; 17:271–2.

31. Wijnen J, de Leeuw W, Vasen H, et al. Familial endometrial cancer in female carriers of *MSH6* germline mutations. Nat Genet 1999; 23:142–4.

32. Park YJ, Shin K-H, Park J-G. Risk of gastric cancer in hereditary nonpolyposis colorectal cancer in Korea. Clin Cancer Res 2000; 6:2994–8.

33. Burke C. Risk stratification for periampullary carcinoma in patients with familial adenomatous polyposis: does Theodore know what to do now? Gastroenterology 2001; 121:1246–8.

34. Björk J, Akerbrant H, Iselius L, et al. Periampullary adenomas and adenocarcinomas in familial adenomatous polyposis: cumulative risks and APC gene mutations. Gastroenterology 2001; 121:1127–35.

35. Lynch HT, Fitzgibbons R Jr, Chong S, et al. Use of doxorubicin and dacarbazine for the management of unresectable intra-abdominal desmoid tumors in Gardner's syndrome. Dis Colon Rectum 1994; 37:260–7.

36. Lynch HT, Fitzgibbons R Jr. Surgery, desmoid tumors, and familial adenomatous polyposis: case report and literature review. Am J Gastroenterol 1996; 91:2598–601.

37. Lynch HT, Lynch PM. Negative genetic test result in familial adenomatous polyposis. Dis Colon Rectum 1999; 42:310–2.

38. Lynch HT, Smyrk TC. Classification of familial adenomatous polyposis: a diagnostic nightmare. Am J Hum Genet 1998; 62:1288–9.

39. Lynch HT, Tinley ST, Lynch J, Vanderhoof J, Lemon SJ. Familial adenomatous polyposis: discovery of a family and its management in a cancer genetics clinic. Cancer 1997; 80:Suppl:614–20.

40. Lynch HT, Smyrk T, Lynch J, Lanspa S, McGinn T, Cavalieri RJ. Genetic counseling in an extended attenuated familial adenomatous polyposis kindred. Am J Gastroenterol 1996; 91:455–9.

41. Lynch HT. Desmoid tumors: genotype-phenotype differences in familial adenomatous polyposis—a nosological dilemma. Am J Hum Genet 1996; 59:1184–5.

42. Brensinger JD, Laken SJ, Luce MC, et al. Variable phenotype of familial adenomatous polyposis in pedigrees with 3′ mutation in the APC gene. Gut 1998; 43:548–52.

43. Laken SJ, Petersen GM, Gruber SB, et al. Familial colorectal cancer in Ashkenazim due to a hypermutable tract in *APC*. Nat Genet 1997; 17:79–83.

44. Prior TW, Chadwick RB, Papp AC, et al. The I1307K polymorphism of the *APC* gene in colorectal cancer. Gastroenterology 1999; 116:58–63.

45. Rozen P, Shomrat R, Strul H, et al. Prevalence of the I1307K *APC* gene variant in Israeli Jews of differing ethnic origin and risk for colorectal cancer. Gastroenterology 1999; 116:54–7.

46. Kinzler KW, Vogelstein B. Lessons from hereditary colon cancer. Cell 1996; 87:159–70.

47. Herrera L, ed. Familial adenomatous polyposis. New York: Alan R. Liss, 1990.

48. Steinbach G, Lynch PM, Phillips RKS, et al. The effect of celecoxib, a cyclooxygenase-2 inhibitor, in familial adenomatous polyposis. N Engl J Med 2000; 342:1946–52.

49. Tonelli F, Valanzano R, Messerini L, Ficari F. Long-term treatment with sulindac in familial adenomatous polyposis: is there an actual efficacy in prevention of rectal cancer? J Surg Oncol 2000; 74:15–20.

50. Burn J, Chapman PD, Mathers J, et al. The protocol for a European double-blind trial of aspirin and resistant starch in familial adenomatous polyposis: the CAPP study. Eur J Cancer 1995; 31A:1385–6.

51. Williamson SLH, Kartheuser A, Coaker J, et al. Intestinal tumorigenesis in the *Apc*1638N mouse treated with aspirin and resistant starch for up to 5 months. Carcinogenesis 1999; 20:805–10.

52. Lynch HT, Thorson AG, Smyrk T. Rectal cancer after prolonged sulindac chemoprevention: a case report. Cancer 1995; 75:936–8.

53. Aarnio M, Sankila R, Pukkala E, et al. Cancer risk in mutation carriers of DNA-mismatch-repair genes. Int J Cancer 1999; 81:214–8.

54. Fusaro RM, Lemon SJ, Lynch HT. The Muir-Torre syndrome: a variant of the hereditary nonpolyposis colorectal cancer syndrome. J Tumor Marker Oncol 1996; 11:19–31.

55. Lynch HT, Richardson JD, Amin M, et al. Variable gastrointestinal and urologic cancers in a Lynch syndrome II kindred. Dis Colon Rectum 1991; 34:891–5.

56. Vasen HFA, Mecklin J-P, Khan PM, Lynch HT. The International Collaborative Group on Hereditary Non-Polyposis Colorectal Cancer (ICG-HNPCC). Dis Colon Rectum 1991; 34:424–5.

57. Vasen HFA, Watson P, Mecklin J-P, Lynch HT. New clinical criteria for hereditary nonpolyposis colorectal cancer (HNPCC, Lynch syndrome) proposed by the International Collaborative Group on HNPCC. Gastroenterology 1999; 116:1453–6.

58. Smyrk TC, Watson P, Kaul K, Lynch HT. Tumor-infiltrating lymphocytes are a marker for microsatellite instability in colorectal carcinoma. Cancer 2001; 91:2417–22.

59. Alexander J, Watanabe T, Wu T-T, Rashid A, Li S, Hamilton SR. Histopathological identification of colon cancer with microsatellite instability. Am J Pathol 2001; 158:527–35.

60. Jass JR, Do K-A, Simms LA, et al. Morphology of sporadic colorectal cancer with DNA replication errors. Gut 1998; 42:673–9.

61. Jass JR, Stewart SM. Evolution of hereditary non-polyposis colorectal cancer. Gut 1992; 33:783–6.

62. Lynch J. The genetics and natural history of hereditary colon cancer. Semin Oncol Nurs 1997; 13:91–8.

63. Nyström-Lahti M, Wu Y, Moisio A-L, et al. DNA mismatch repair gene mutations in 55 kindreds with verified or putative hereditary non-polyposis colorectal cancer. Hum Mol Genet 1996; 5:763–9.

64. Wu Y, Berends MJW, Sijmons RH, et al. A role for MLH3 in hereditary nonpolyposis colorectal cancer. Nat Genet 2001; 29:137–8.

65. Lindor NM, Burgart LJ, Leontovich O, et al. Immunohistochemistry versus microsatellite instability testing in phenotyping colorectal tumors. J Clin Oncol 2002; 20:1043–8.

66. Yan H, Papadopoulos N, Marra G, et al. Conversion of diploidy to haploidy. Nature 2000; 403:723–4.

67. Nakagawa H, Yan H, Lockman J, et al. Allele separation facilitates interpretation of potential splicing alterations and genomic rearrangements. Cancer Res 2002; 62:4579–82.

68. Wijnen J, van der Klift H, Vasen H, et al. *MSH2* genomic deletions are a frequent cause of HNPCC. Nat Genet 1998; 20:326–8.

69. Burke W, Petersen G, Lynch P, et al. Recommendations for follow-up care of individuals with an inherited predisposition to cancer. I. Hereditary nonpolyposis colon cancer. JAMA 1997; 277:915–9.

70. Järvinen HJ, Aarnio M, Mustonen H, et al. Controlled 15-year trial on screening for colorectal cancer in families with hereditary nonpolyposis colorectal cancer. Gastroenterology 2000; 118:829–34.

71. Lynch HT. Is there a role for prophylactic subtotal colectomy among hereditary nonpolyposis colorectal cancer germline mutation carriers? Dis Colon Rectum 1996; 39:109–10.

72. Houlston RS, Collins A, Slack J, Morton NE. Dominant genes for colorectal cancer are not rare. Ann Hum Genet 1992; 56:99–103.

73. Ramsey SD, Clarke L, Etzioni R, Higashi M, Berry K, Urban N. Cost-effectiveness of microsatellite instability screening as a method for detecting hereditary nonpolyposis colorectal cancer. Ann Intern Med 2001; 135:577–88.

74. Rodriguez-Bigas MA, Boland CR, Hamilton SR, et al. A National Cancer Institute workshop on hereditary nonpolyposis colorectal cancer syndrome: meeting highlights and Bethesda guidelines. J Natl Cancer Inst 1997; 89:1758–62.

75. Fearon ER, Vogelstein B. A genetic model for colorectal tumorigenesis. Cell 1990; 61:759–67.

76. Ilyas M, Straub J, Tomlinson IPM, Bodmer WF. Genetic pathways in colorectal and other cancers. Eur J Cancer 1999; 35:335–51.

77. Konishi M, Kikuchi-Yanoshita R, Tanaka K, et al. Molecular nature of colon tumors in hereditary nonpolyposis colorectal cancer, familial polyposis, and sporadic colon cancer. Gastroenterology 1996; 111:307–17.

78. Huang J, Papadopoulos N, McKinley AJ, et al. APC mutations in colorectal tumors with mismatch repair deficiency. Proc Natl Acad Sci U S A 1996; 93:9049–54.

79. Duval A, Rolland S, Compoint A, et al. Evolution of instability at coding and non-coding repeat sequences in human MSI-H colorectal cancers. Hum Mol Genet 2001; 10:513–8.

80. Chadwick RB, Jiang G-L, Bennington GA, et al. Candidate tumor suppressor RIZ is frequently involved in colorectal carcinogenesis. Proc Natl Acad Sci U S A 2000; 97:2662–7.

81. Chadwick RB, Pyatt RE, Niemann TH, et al. Hereditary and somatic DNA mismatch repair gene mutations in sporadic endometrial carcinoma. J Med Genet 2001; 38:461–6.

82. Malkhosyan S, Rampino N, Yamamoto H, Perucho M. Frameshift mutator mutations. Nature 1996; 382:499–500.

83. Kane MF, Loda M, Gaida GM, et al. Methylation of the *hMLH1* promoter correlates with lack of expression of hMLH1 in sporadic colon tumors and mismatch repair-defective human tumor cell lines. Cancer Res 1997; 57:808–11.

84. Herman JG, Umar A, Polyak K, et al. Incidence and functional consequences of *hMLH1* promoter hypermethylation in colorectal carcinoma. Proc Natl Acad Sci U S A 1998; 95:6870–5.

85. Cunningham JM, Christensen ER, Tester DJ, et al. Hypermethylation of the hMLH1 promoter in colon cancer with microsatellite instability. Cancer Res 1998; 58:3455–60.

86. Veigl ML, Kasturi L, Olechnowicz J, et al. Biallelic inactivation of hMLH1 by epigenetic gene silencing, a novel mechanism causing human MSI cancers. Proc Natl Acad Sci U S A 1998; 95:8698–702.

87. Nakagawa H, Chadwick RB, Peltomäki P, Plass C, Nakamura Y, de la Chapelle A. Loss of imprinting of the insulin-like growth factor II gene occurs by biallelic methylation in a core region of *H19*-associated CTCF-binding sites in colorectal cancer. Proc Natl Acad Sci U S A 2001; 98:591–6.

88. Hemminki A, Markie D, Tomlinson I, et al. A serine/threonine kinase gene defective in Peutz-Jeghers syndrome. Nature 1998; 391:184–7.

89. Howe JR, Roth S, Ringold JC, et al. Mutations in the SMAD4/DPC4 gene in juvenile polyposis. Science 1998; 280:1086–8.

90. Howe JR, Bair JL, Sayed MG, et al. Germline mutations of the gene encoding bone morphogenetic

protein receptor 1A in juvenile polyposis. Nat Genet 2001; 28:184–7.

91. Lynch ED, Ostermeyer EA, Lee MK, et al. Inherited mutations in PTEN that are associated with breast cancer, Cowden disease, and juvenile polyposis. Am J Hum Genet 1997; 61:1254–60.

92. Liaw D, Marsh DJ, Li J, et al. Germline mutations of the *PTEN* gene in Cowden disease, an inherited breast and thyroid cancer syndrome. Nat Genet 1997; 16:64–7.

93. Marsh DJ, Coulon V, Lunetta KL, et al. Mutation spectrum and genotype-phenotype analyses in Cowden disease and Bannayan-Zonana syndrome, two hamartoma syndromes with germline PTEN mutation. Hum Mol Genet 1998; 7:507–15.

94. Eng C, Ji H. Molecular classification of the inherited hamartoma polyposis syndromes: clearing the muddied waters. Am J Hum Genet 1998; 62:1020–2.

95. Matloff ET, Shappell H, Brierley K, Bernhardt BA, McKinnon W, Peshkin BN. What would you do? Specialists' perspectives on cancer genetic testing, prophylactic surgery, and insurance discrimination. J Clin Oncol 2000; 18:2484–92.

96. Pokorski RJ, Sanderson P, Bennett N, et al. Insurance issues and genetic testing: challenges and recommendations: workshop no. 1. Cancer 1997; 80:Suppl:627.

97. The Ad Hoc Committee on Genetic Testing/Insurance Issues. Genetic testing and insurance. Am J Hum Genet 1995; 56:327–31.

98. Balanced Budget Act of 1997, Pub. L. No. 105–33, 105th Cong. 1st Sess.

99. Druker BJ, Talpaz M, Resta DJ, et al. Efficacy and safety of a specific inhibitor of the BCR-ABL tyrosine kinase in chronic myeloid leukemia. N Engl J Med 2001; 344:1031–7.

100. Druker BJ, Sawyers CL, Kantarjian H, et al. Activity of a specific inhibitor of the BCR-ABL tyrosine kinase in the blast crisis of chronic myeloid leukemia and acute lymphoblastic leukemia with the Philadelphia chromosome. N Engl J Med 2001; 344:1038–42. [Erratum, N Engl J Med 2001; 345:232.]

101. Al-Tassan N, Chmiel NH, Maynard J, et al. Inherited variants of MYH associated with somatic G:C → T:A mutations in colorectal tumors. Nat Genet 2002; 30:227–32.

102. Sieber OM, Lipton L, Crabtree M, et al. Multiple colorectal adenomas, classic adenomatous polyposis, and germ-line mutations in *MYH*. N Engl J Med 2003; 348:791–9.

9

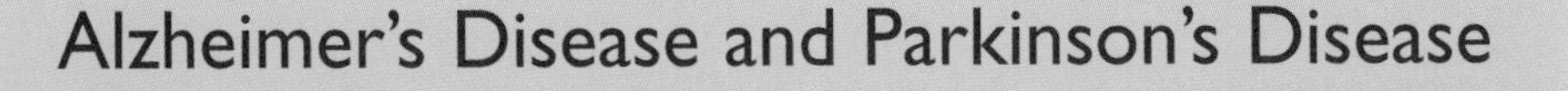

Alzheimer's Disease and Parkinson's Disease

ROBERT L. NUSSBAUM, M.D., and CHRISTOPHER E. ELLIS, Ph.D.

The incidence of many common diseases is increased among the relatives of affected patients, but the pattern of inheritance rarely follows Mendel's laws. Instead, such common diseases are thought to result from a complex interaction among multiple predisposing genes and other factors, including environmental contributions and chance occurrences. Identifying the genetic contribution to such complex diseases is a major challenge for genomic medicine. However, as so clearly foreseen nearly 350 years ago by the English physiologist William Harvey,[1] finding the genetic basis for rarer, mendelian forms of a disease may illuminate the etiologic process and pathogenesis of the more common, complex forms. This is illustrated in the progress made in understanding Alzheimer's disease and Parkinson's disease through the investigation of the rare, clearly inherited forms of these diseases. The molecular basis of neurodegenerative disorders was reviewed in the *New England Journal of Medicine* in 1999.[2]

ALZHEIMER'S DISEASE

The most common neurodegenerative disease, Alzheimer's disease constitutes about two thirds of cases of dementia overall (ranging in various studies from 42 to 81 percent of all dementias), with vascular causes and other neurodegenerative diseases such as Pick's disease and diffuse Lewy-body dementia making up the majority of the remaining cases.[3,4]

Alzheimer's disease is a progressive neurologic disease that results in the irreversible loss of neurons, particularly in the cortex and hippocampus.[5] The clinical hallmarks are progressive impairment in memory, judgment, decision making, orientation to physical surroundings, and language. Diagnosis is based on neurologic examination and the exclusion of other causes of dementia; a definitive diagnosis can be made only at autopsy. The pathological hallmarks are neuronal loss, extracellular senile plaques containing the peptide

Originally published April 3, 2003

β amyloid, and neurofibrillary tangles; the latter are composed of a hyperphosphorylated form of the microtubular protein tau.[6] Amyloid in senile plaques is the product of cleavage of a much larger protein, the β-amyloid precursor protein, by a series of proteases, the α-, β-, and γ-secretases.[7] The γ-secretase, in particular, appears to be responsible for generating one particular β-amyloid peptide—$A\beta_{42}$—that is 42 amino acids in length and has pathogenetic importance, because it can form insoluble toxic fibrils and accumulates in the senile plaques isolated from the brains of patients with Alzheimer's disease.[8,9]

Measures of the prevalence of Alzheimer's disease differ depending on the diagnostic criteria used, the age of the population surveyed, and other factors, including geography and ethnicity.[10,11] Excluding persons with clinically questionable dementia, Alzheimer's disease has a prevalence of approximately 1 percent among those 65 to 69 years of age and increases with age to 40 to 50 percent among persons 95 years of age and over[10] (Fig. 9.1). Although the mean age at the onset of dementia is approximately 80 years,[3] early-onset disease, defined arbitrarily and variously as the illness occurring before the age of 60 to 65 years, can occur but is rare. In one community-based study in France, the prevalence of early-onset disease (defined by an onset before the age of 61 years) was 41 per 100,000; thus, early-onset cases make up about 6 to 7 percent of all cases of Alzheimer's disease.[12] About 7 percent of early-onset cases are familial, with an autosomal dominant pattern of inheritance and high penetrance.[12] Thus, familial forms of early-onset Alzheimer's disease, inherited in an autosomal dominant manner, are rare; however, their importance extends far beyond their frequency, because they have allowed researchers to identify some of the critical pathogenetic pathways of the disease.

Missense mutations that alter a single amino acid and therefore gene function have been identified in three genes in families with early-onset autosomal dominant Alzheimer's disease. Family linkage studies and DNA sequencing identified mutations responsible for early-onset autosomal dominant forms of the disease in the gene encoding β-amyloid precursor protein itself on chromosome 21 (Fig. 9.2), as well as in two genes with similarity to

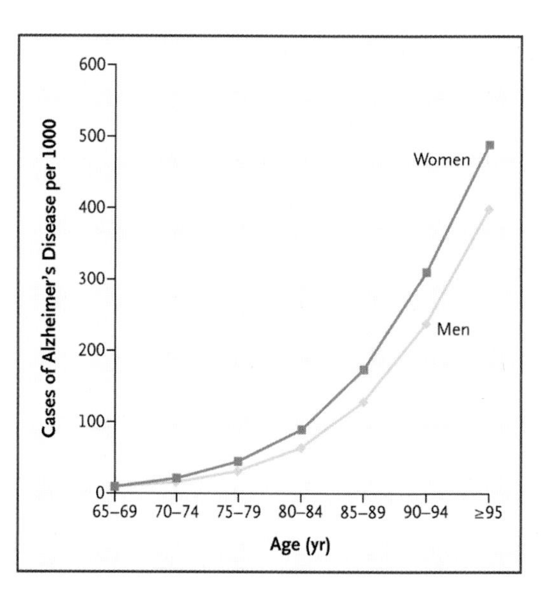

Figure 9.1 Prevalence of Alzheimer's Disease as a Function of Age in Men and Women.

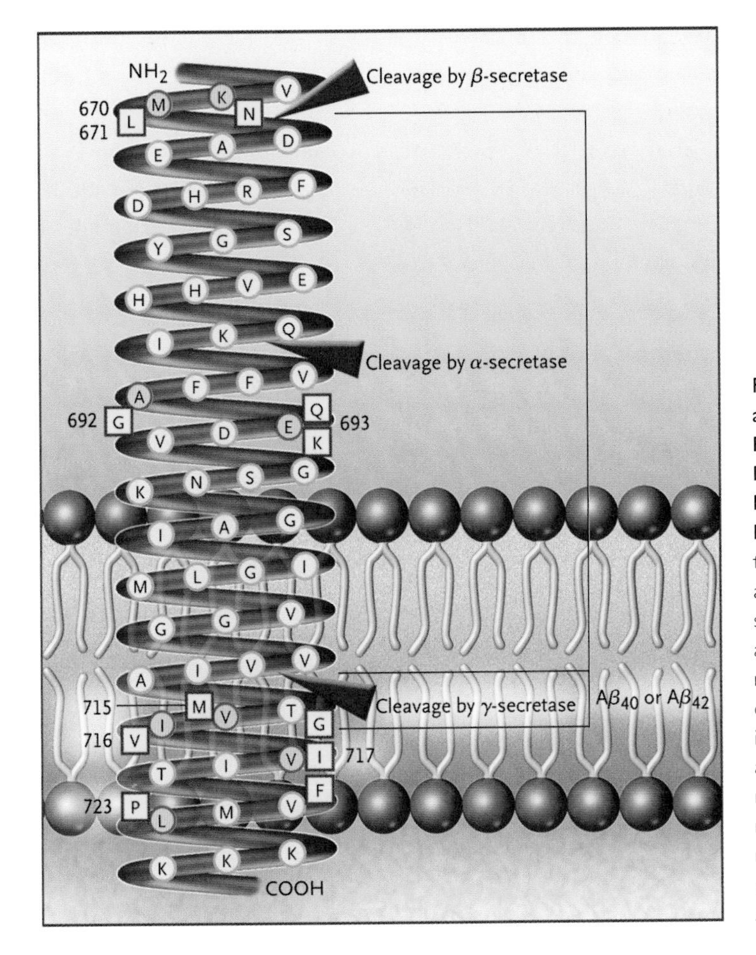

Figure 9.2 Altered Amino Acid Residues in a Segment of the β-Amyloid Precursor Protein Adjacent to Its Transmembrane Domain Resulting from Missense Mutations and Causing Early-Onset Familial Alzheimer's Disease. Letters are the single-letter code for amino acids in β-amyloid precursor protein, and numbers show the position of the affected amino acid. The altered amino acid residues are near the sites of β-, α-, and γ-secretase cleavage (red triangles). Normal residues involved in missense mutations are shown as green circles, whereas the amino acid residues representing various missense mutations are shown as yellow boxes. The mutations lead to the accumulation of toxic peptide $A\beta_{42}$ rather than the wild-type $A\beta_{40}$ peptide.

each other, presenilin 1 (*PSEN1*) on chromosome 14 and presenilin 2 (*PSEN2*) on chromosome 1. *PSEN1* mutations are more common than *PSEN2* mutations. In a study of French families, for example, half of patients with familial, early-onset Alzheimer's disease that was inherited as an autosomal dominant trait had mutations in *PSEN1*, whereas approximately 16 percent of families had mutations in the β-amyloid precursor protein (*βAPP*) gene itself.[12] *PSEN2* mutations were not found, and the genes responsible for the remaining 30 percent or so of cases were unknown.

The presenilin and *βAPP* mutations found in patients with familial early-onset Alzheimer's disease appear to result in the increased production of $A\beta_{42}$, which is probably the primary neurotoxic species involved in the pathogenesis of the disease[7,13] (Fig. 9.3). In these forms of Alzheimer's disease, mutations in *βAPP* itself or in the presenilins can shift the cleavage site to favor the γ-secretase site[14] and, in particular, to favor increased production of the toxic $A\beta_{42}$ peptide over the shorter, less toxic $A\beta_{40}$ peptide. Presenilin 1 may in fact be the γ-secretase itself or a necessary cofactor in γ-secretase activity.[15] The toxic peptide is increased in the serum of patients with various *βAPP*, *PSEN1*, and *PSEN2* mutations causing early-onset Alzheimer's disease.[16] Cultured cells transfected in order to express the normal β-amyloid precursor protein generally process approximately 10 percent of the

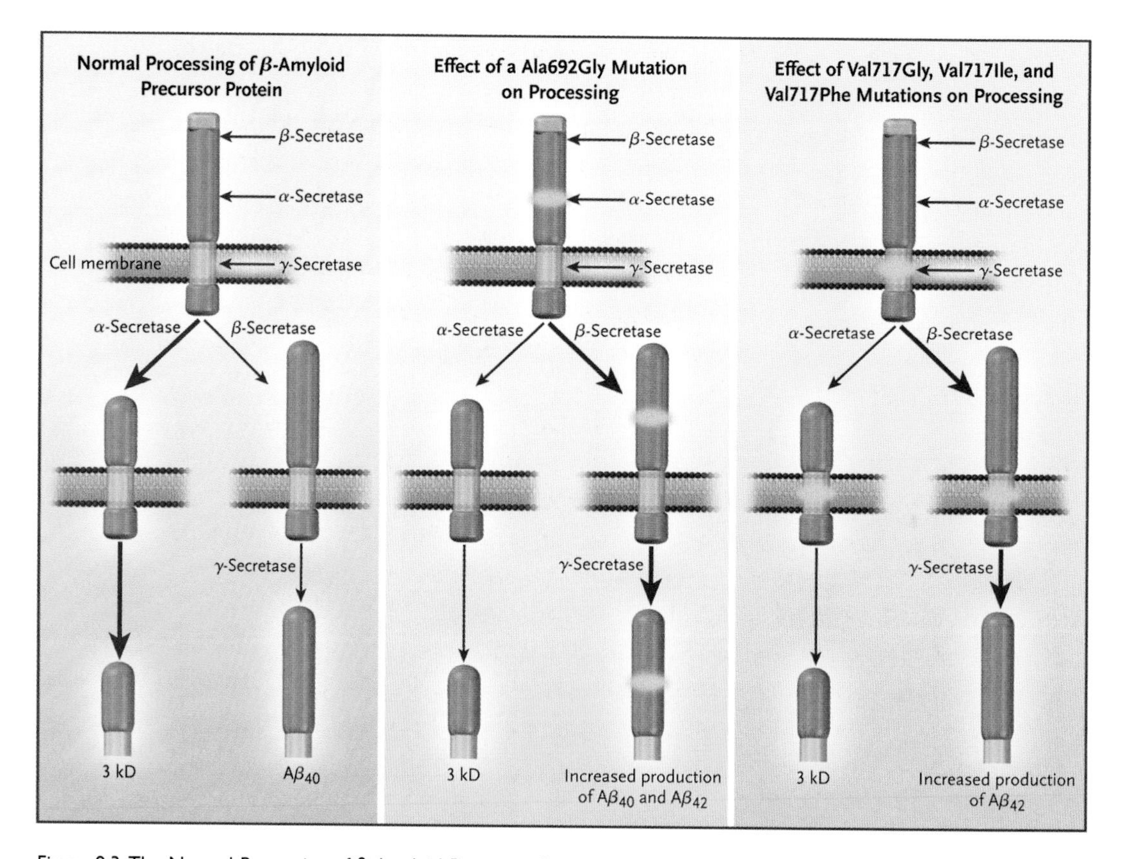

Figure 9.3 The Normal Processing of β-Amyloid Precursor Protein as Well as the Effect on Processing of Alterations in the Protein Resulting from Missense Mutations Associated with Early-Onset Familial Alzheimer's Disease. These mutations (indicated by the yellow burst symbols) either interfere with α-secretase or enhance β- or γ-secretase cleavage, resulting in an increase in the production of the toxic $A\beta_{42}$ peptide rather than the wild-type $A\beta_{40}$ peptide. Drugs with selectivity for certain secretases might reduce or eliminate the processing of β-amyloid precursor protein to the toxic $A\beta_{42}$ peptide and therefore help prevent Alzheimer's disease or slow its progression. The thickness of the arrows represents the amount of each peptide being made relative to the other peptides.

protein into the toxic $A\beta_{42}$ peptide. Expression of various mutant *βAPP* or *PSEN1* genes associated with early-onset familial Alzheimer's disease can result in an increase in the production of $A\beta_{42}$ by a factor of up to 10.[17–19] The identification of mutations in *βAPP* and the presenilins in early-onset familial Alzheimer's disease not only suggests a common mechanism through which mutations in these genes may exert their deleterious effects (i.e., increased production or decreased clearance of $A\beta_{42}$ and formation of a protein aggregate, the amyloid plaque) but also provides evidence of a direct role of the $A\beta_{42}$ peptide and presenilins in the pathogenesis of the disease.[20] In contrast, mutations in the tau gene, which encodes a protein contained within another neuropathologic structure in Alzheimer's disease, the neurofibrillary tangle, have not been identified in families with hereditary Alzheimer's disease, although they are seen in another, rarer neurodegenerative disorder, frontotemporal degeneration with parkinsonism[21,22] (Fig. 9.3).

As important as the rare familial early-onset forms of Alzheimer's disease have been for understanding the pathogenesis of the disease, the majority of patients of any age have sporadic (nonfamilial) disease in which no mutation in the *βAPP* or presenilin genes has been identified. However, another genetic risk factor, variants of *APOE*, the gene that encodes apolipoprotein E, a constituent of the low-density lipoprotein particle, has been associated with Alzheimer's disease.[23] Three variants of the gene and the protein are found in human populations and result from changes in single amino acids in apolipoprotein E (referred to as the *APOE* ϵ2, ϵ3, and ϵ4 alleles). Carrying one *APOE* ϵ4 allele nearly doubles the lifetime risk of Alzheimer's disease (from 15 percent to 29 percent), whereas not carrying an *APOE* ϵ4 allele cuts the risk by 40 percent.[24] Initially, survival curves analyzing the effect of the *APOE* ϵ4 allele on the occurrence of Alzheimer's disease suggested that 70 to 90 percent of persons without this allele were disease-free at the age of 80 years, whereas 30 to 60 percent of those with one *APOE* ϵ4 allele and only 10 percent of homozygous persons surviving to the age of 80 were disease-free.[23] A more recent study also provided evidence that *APOE* ϵ4 has a role in Alzheimer's disease, but the effect was less marked, with the rate of disease-free survival as high as 70 percent in homozygous persons.[25]

Although the magnitude of the effect of the *APOE* ϵ4 allele differs among studies, there appears to be a dose effect, in that disease-free survival was lower in homozygous persons than in heterozygous persons. This observation has led to the conclusion that the primary effect of the *APOE* ϵ4 allele is to shift the age at onset an average of approximately 5 to 10 years earlier in the presence of one allele and 10 to 20 years earlier in the presence of two alleles in persons with an underlying susceptibility to Alzheimer's disease.[26] The molecular mechanisms by which the various *APOE* alleles alter the age at onset and, therefore, the lifetime risk of Alzheimer's disease are unknown. A number of associations of the disease with variants of genes other than *APOE* have also been reported but remain to be confirmed and are the subject of ongoing research.[27]

Because of the relative rarity of *βAPP*, *PSEN1*, and *PSEN2* mutations in patients with late-onset Alzheimer's disease, we believe that molecular testing for mutations in these genes should be restricted to those with an elevated probability of having such mutations—that is, persons with early-onset disease or a family history of the disease. At-risk, symptomatic relatives of persons with documented mutations in *βAPP* or one of the presenilins may also request testing for the purposes of family, financial, or personal planning. Testing of a presymptomatic person should be undertaken with extreme care and only after extensive pretest counseling, so that the person requesting the test is aware of the potential for severe psychological complications of testing positive for an incurable, devastating illness. There may also be serious ramifications in the area of employment and in obtaining life, long-term care, disability, or health insurance. Also important is that a positive test may indicate that other family members, who may not have participated in any counseling or consented to testing, will be identified as being at a substantially increased risk for early-onset Alzheimer's disease by virtue of their relationship to the person who tests positive.

The usefulness of testing for the *APOE* ϵ4 allele is also limited. Finding one or two *APOE* ϵ4 alleles in a symptomatic person with dementia certainly increases the likelihood that one is dealing with Alzheimer's disease and might be used as an adjunct to clinical diagnosis.[28] On the other hand, since 50 percent of patients with autopsy-proved Alzheimer's disease did not carry an *APOE* ϵ4 allele, its negative predictive value in a symptomatic person is very limited.[24] *APOE* ϵ4 testing in asymptomatic persons has very poor positive and negative predictive values and should not be used.[24]

Insights derived from the identification of mutations in rare families with early-onset Alzheimer's disease are proving useful for identifying therapeutic targets and creating animal models for evaluating therapies.[29] For example, β-secretase inhibitors have been developed and may prove useful in treating Alzheimer's disease by reducing $A\beta_{42}$ production.[30] Transgenic mice expressing mutant β-amyloid precursor protein have an age-dependent increase in the amount of $A\beta_{42}$ formation, increased plaque formation, and spatial memory deficits; they have, however, only a minimal loss of neurons.[31] In addition, mice transgenic for both a *βAPP* and a *PSEN1* mutation show accelerated deposition of $A\beta_{42}$, as compared with mice expressing either transgene alone.[32] In transgenic mice with a mutant β-amyloid precursor protein, immunization with $A\beta_{42}$ resulted in a decrease in plaque formation and an amelioration of memory loss.[32–34] However, phase 2 clinical trials investigating immunization therapy with $A\beta_{42}$[35] had to be suspended because of an increased risk of aseptic meningoencephalitis.[35–37] In addition, other drugs such as statins, clioquinol, and certain nonsteroidal antiinflammatory drugs[38] are being evaluated in mouse models of these rare, heritable forms of Alzheimer's disease.

PARKINSON'S DISEASE

Parkinson's disease is the second most common neurodegenerative disorder, after Alzheimer's disease. It is characterized clinically by parkinsonism (resting tremor, bradykinesia, rigidity, and postural instability)[39] and pathologically by the loss of neurons in the substantia nigra and elsewhere in association with the presence of ubiquinated protein deposits in the cytoplasm of neurons (Lewy bodies)[40,41] and thread-like proteinaceous inclusions within neurites (Lewy neurites). Parkinson's disease has a prevalence of approximately 0.5 to 1 percent among persons 65 to 69 years of age, rising to 1 to 3 percent among persons 80 years of age and older.[42] The diagnosis is made clinically, although other disorders with prominent symptoms and signs of parkinsonism, such as postencephalitic, drug-induced, and arteriosclerotic parkinsonism, may be confused with Parkinson's disease until the diagnosis is confirmed at autopsy.[43]

A genetic component in Parkinson's disease was long thought to be unlikely, because most patients had sporadic disease and initial studies of twins showed equally low rates of concordance in monozygotic and dizygotic twins.[44] The view that genetics was involved in some forms of Parkinson's disease was strengthened, however, by the observation that monozygotic twins with an onset of disease before the age of 50 years do have a very high rate of concordance—much higher than that of dizygotic twins with early-onset disease.[44,45] Furthermore, regardless of the age at onset, the apparent rate of concordance among monozygotic twins can be significantly increased if abnormal striatial dopaminergic uptake in the asymptomatic twin of a discordant pair, as revealed by positron-emission tomography with fluorodopa F18, is used as a sign of presymptomatic Parkinson's disease.[46,47] An increased risk of Parkinson's disease was also seen among the first-degree relatives of patients,[48,49] particularly when the results of positron-emission tomography of asymptomatic relatives were taken into account,[50] providing further evidence of the existence of a genetic component to the disease.

However, the real advance occurred when a small number of families with early-onset, Lewy-body–positive autosomal dominant Parkinson's disease were identified.[51] Investigation of these families, of Mediterranean and German origin, led to the identification of two missense mutations (Ala53Thr and Ala30Pro) in the gene encoding α-synuclein, a small presynaptic protein of unknown function.[52,53] Although mutations in α-synuclein have proved to be extremely rare in patients with Parkinson's disease, they did provide the first clue that this protein could be involved in the molecular chain of events leading to the disease. The importance of α-synuclein was greatly enhanced by the discovery that the Lewy bodies and Lewy neurites found in Parkinson's disease in general contain aggregates of α-synuclein[54,55] (Fig. 9.4). Molecules of α-synuclein protein are prone to form into oligomers in vitro; proteins carrying the missense mutations Ala53Thr and Ala30Pro seem to be even more prone to do so.[56]

Although the study of families with early-onset Parkinson's disease proves that abnormal α-synuclein can cause the disease, it is still unclear whether fibrils of aggregated α-synuclein, as seen in Lewy bodies and Lewy neurites, have a critical causative role in the more common forms of Parkinson's disease or are simply a marker for the underlying pathogenetic process. Lewy bodies positive for α-synuclein are found not only in various subnuclei of the substantia nigra, the locus ceruleus, and other brain-stem and thalamic nuclei of patients with Parkinson's disease, but also in a more diffuse distribution, including the cortex in some patients with Parkinson's disease as well as in patients with dementia of the diffuse Lewy-body type.[57,58] Aggregated α-synuclein in glia is also a feature of multiple-system atrophy,[59] leading to the coining of a new nosologic term, "synucleinopathy," to refer to the class of neurodegenerative diseases associated with aggregated α-synuclein.[60]

Autosomal recessive juvenile parkinsonism is another genetic neurologic syndrome that has provided important insights into Parkinson's disease. Autosomal recessive juvenile parkinsonism is a relatively rare syndrome that shares many of the features of parkinsonism, including responsiveness to levodopa and loss of nigrostriatal and locus ceruleus neurons, but it has a very early onset (before the age of 40 years), a slow clinical course extending over many decades, and no Lewy bodies or Lewy neurites at autopsy.[61] Genetic mapping of the syndrome to 6q25–27 led to the identification of mutations responsible for autosomal recessive juvenile parkinsonism in a gene encoding a protein termed parkin.[62] Parkin is expressed primarily in the nervous system and is one member of a family of proteins known as E3 ubiquitin ligases, which attach short ubiquitin peptide chains to proteins, a process termed ubiquination, thereby tagging them for degradation through the proteosomal degradation pathway.

Autosomal recessive juvenile parkinsonism results from a loss of function of both copies of the parkin gene,[63–65] leading to autosomal recessive inheritance, as opposed to the missense mutations that alter α-synuclein and cause a dominantly inherited disorder. More recently, however, the spectrum of disease known to be caused by parkin mutations has

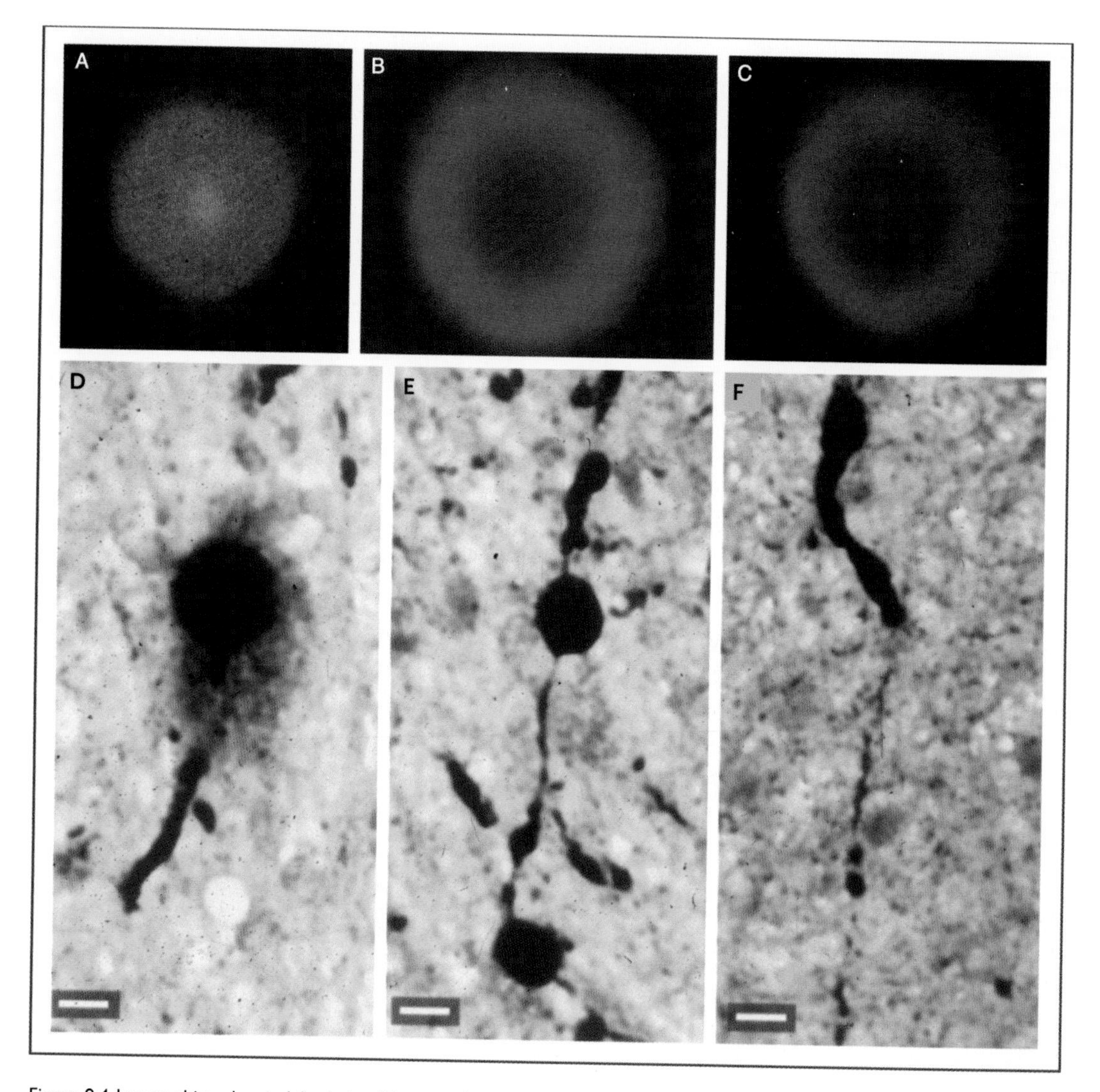

Figure 9.4 Immunohistochemical Analysis of Sections from the Substantia Nigra of a Patient with Sporadic Parkinson's Disease, Indicating the Involvement of α-Synuclein in the Formation of Lewy Bodies and Lewy Neurites. Panel A shows a Lewy body stained with antibody against ubiquitin (green) (×3000). Panel B shows the same Lewy body stained with antibody against α-synuclein (red) (×3000). Panel C, which merges the images shown in Panels A and B, shows that Lewy bodies contain a central core of ubiquinated proteins and α-synuclein surrounded by a rim of α-synuclein–positive fibrillar material (×3000). Panels D, E, and F show neuronal processes from the substantia nigra of a patient with sporadic Parkinson's disease in which neurites are ballooned and dilated and stain for α-synuclein (black stain). Scale bars in Panels D, E, and F indicate 10 μm. (Adapted from Mezey et al.[55])

broadened, with apparently sporadic Parkinson's disease occurring in adulthood, as late as in the fifth and sixth decades of life, in association with parkin gene mutations.[66] There have even been a few patients with apparently classic sporadic Parkinson's disease with an onset in adulthood who appear to have only one mutant parkin allele, although an exhaustive demonstration that the other allele is truly normal and not harboring an unusual mutation outside the coding sequence and its immediate environs is still lacking. Precisely what role parkin mutations have in the majority of cases of Parkinson's disease and whether the heterozygous state (which is far more common in the population than is homozygosity for two mutant alleles) represents an important risk factor remain to be established.

Recent evidence suggests that ubiquination by parkin may be important in the normal turnover of α-synuclein.[67] Also of interest is the finding in one family of a few members with Parkinson's disease who had a deleterious missense mutation in the gene encoding a neuron-specific C-terminal ubiquitin hydrolase, another gene involved in ubiquitin metabolism.[68] The obvious inference from these disparate pieces of data is that aggregation of abnormal proteins, dysfunctional ubiquitin-mediated degradation machinery, or both may be important steps in the pathogenesis of Parkinson's disease.

In addition to the α-synuclein, parkin, and ubiquitin C-hydrolase genes, at least five other loci have been proposed for autosomal dominant[69–71] and autosomal recessive[72–74] Parkinson's disease (Table 9.1). Genetic analysis of the more common, sporadic forms of Parkinson's disease suggests that there is a component of heritability in the forms that are not clearly inherited as autosomal dominant or recessive traits.[75–78] For example, certain alleles at a complex DNA-repeat polymorphic locus approximately 10 kilobase pairs upstream of the α-synuclein gene have been shown to be associated with sporadic Parkinson's disease in some populations, but not in others.[79–82] Positive identification of the genes at these loci is likely to provide additional genes and proteins that can be studied for their roles in the pathogenesis of the disease.

Because of the extreme rarity of α-synuclein mutations, genetic testing for these mutations should be performed only on a research basis when a strong family history of autosomal dominant Parkinson's disease is present. Homozygous parkin mutations are found in the

Table 9.1 Mutations in Single Genes That Lead to Parkinson's Disease

Locus	Gene	Location	Mode of Inheritance	Where Found
PARK1	α-Synuclein	4q21	Autosomal dominant	Greece, Italy, and Germany
PARK2	Parkin	6q25–27	Autosomal recessive; may also be autosomal dominant	Ubiquitous
PARK3	Unknown	2p13	Autosomal dominant	Germany
PARK4	Unknown	4p15	Autosomal dominant	United States
PARK5	Ubiquitin C-terminal hydrolase	4p14	May be autosomal dominant	Germany
PARK6	Unknown	1p35	Autosomal recessive	Italy
PARK7	DJ-1	1p36	Autosomal recessive	Netherlands
PARK8	Unknown	12p11.2–q13.1	Autosomal dominant	Japan

nearly half of patients presenting with apparent Parkinson's disease in childhood and adolescence and perhaps 5 percent of young adults with Parkinson's disease.[64] There is little evidence supporting a role for mutations in the parkin gene in typical late-onset Parkinson's disease, and neither α-synuclein nor parkin gene testing is currently available as a routine clinical service.

CONCLUSIONS

The common neurodegenerative diseases are predominantly idiopathic disorders of unknown pathogenesis. As the examples of Alzheimer's disease and Parkinson's disease demonstrate, however, the genetic mapping and gene-isolation tools created by the Human Genome Project over the past decade have greatly accelerated the rate of identification of genes involved in the rare inherited forms of these diseases and are now being used to determine the genetic contributions to the more common, multifactorial forms of these diseases. The emergence of a consensus hypothesis—aggregates of $A\beta_{42}$ and α-synuclein are neurotoxic in Alzheimer's disease and Parkinson's disease, respectively—may explain the pathogenesis not only of the inherited forms of these diseases but also of the idiopathic variety. Such insights into causation and pathogenesis are helping to identify new treatment targets for these debilitating disorders.

ACKNOWLEDGMENT

We are indebted to Dr. John Hardy for helpful discussions and suggestions for figures.

REFERENCES

1. Garrod A. The lessons of rare maladies. Lancet 1928; 1:1055–60.
2. Martin JB. Molecular basis of the neurodegenerative disorders. N Engl J Med 1999; 340:1970–80. [Erratum, N Engl J Med 1999; 341:1407.]
3. Helmer C, Joly P, Letenneur L, Commenges D, Dartigues JF. Mortality with dementia: results from a French prospective community-based cohort. Am J Epidemiol 2001; 154:642–8.
4. Aronson MK, Ooi WL, Geva DL, Masur D, Blau A, Frishman W. Dementia: age-dependent incidence, prevalence, and mortality in the old old. Arch Intern Med 1991; 151:989–92.
5. McKhann G, Drachman D, Folstein M, Katzman R, Price D, Stadlan EM. Clinical diagnosis of Alzheimer's disease: report of the NINCDS-ADRDA Work Group under the auspices of Department of Health and Human Services Task Force on Alzheimer's Disease. Neurology 1984; 34:939–44.
6. Clark CM, Ewbank D, Lee VM-Y, Trojanowski JQ. Molecular pathology of Alzheimer's disease: neuronal cytoskeletal abnormalities. In: Growdon JH, Rossor MN, eds. The dementias. Vol. 19 of Blue books of practical neurology. Boston: Butterworth–Heinemann, 1998:285–304.
7. Hutton M, Perez-Tur J, Hardy J. Genetics of Alzheimer's disease. Essays Biochem 1998; 33: 117–31.
8. Iwatsubo T, Odaka A, Suzuki N, Mizusawa H, Nukina N, Ihara Y. Visualization of A beta 42(43) and A beta 40 in senile plaques with end-specific A beta monoclonals: evidence that an initially deposited species is A beta 42(43). Neuron 1994; 13:45–53.
9. Esler WP, Wolfe MS. A portrait of Alzheimer secretases—new features and familiar faces. Science 2001; 293:1449–54.
10. Hy LX, Keller DM. Prevalence of AD among whites: a summary by levels of severity. Neurology 2000; 55:198–204.
11. Hendrie HC, Ogunniyi A, Hall KS, et al. Incidence of dementia and Alzheimer disease in 2 communities: Yoruba residing in Ibadan, Nigeria, and African Americans residing in Indianapolis, Indiana. JAMA 2001; 285:739–47.

12. Campion D, Dumanchin C, Hannequin D, et al. Early-onset autosomal dominant Alzheimer disease: prevalence, genetic heterogeneity, and mutation spectrum. Am J Hum Genet 1999; 65:664–70.

13. Steiner H, Haass C. Intramembrane proteolysis by presenilins. Nat Rev Mol Cell Biol 2000; 1:217–24.

14. Borchelt DR, Thinakaran G, Eckman CB, et al. Familial Alzheimer's disease-linked presenilin 1 variants elevate Abeta 1–42/1–40 ratio in vitro and in vivo. Neuron 1996; 17:1005–13.

15. Wolfe MS, Xia W, Ostaszewski BL, Diehl TS, Kimberly WT, Selkoe DJ. Two transmembrane aspartates in presenilin-1 required for presenilin endoproteolysis and gamma-secretase activity. Nature 1999; 398:513–7.

16. Scheuner D, Eckman C, Jensen M, et al. Secreted amyloid beta-protein similar to that in the senile plaques of Alzheimer's disease is increased in vivo by the presenilin 1 and 2 and APP mutations linked to familial Alzheimer's disease. Nat Med 1996; 2:864–70.

17. Murayama O, Tomita T, Nihonmatsu N, et al. Enhancement of amyloid beta 42 secretion by 28 different presenilin 1 mutations of familial Alzheimer's disease. Neurosci Lett 1999; 265:61–3.

18. Mehta ND, Refolo LM, Eckman C, et al. Increased Abeta42(43) from cell lines expressing presenilin 1 mutations. Ann Neurol 1998; 43:256–8.

19. Eckman CB, Mehta ND, Crook R, et al. A new pathogenic mutation in the APP gene (I716V) increases the relative proportion of A beta 42(43). Hum Mol Genet 1997; 6:2087–9.

20. Hardy J, Selkoe DJ. The amyloid hypothesis of Alzheimer's disease: progress and problems on the road to therapeutics. Science 2002; 297:353–6. [Erratum, Science 2002; 297:2209.]

21. Lynch T, Sano M, Marder KS, et al. Clinical characteristics of a family with chromosome 17-linked disinhibition-dementia-parkinsonism-amyotrophy complex. Neurology 1994; 44:1878–84.

22. Hutton M, Lendon CL, Rizzu P, et al. Association of missense and 5′-splice-site mutations in tau with the inherited dementia FTDP-17. Nature 1998; 393:702–5.

23. Strittmatter WJ, Roses AD. Apolipoprotein E and Alzheimer's disease. Annu Rev Neurosci 1996; 19:53–77.

24. Seshadri S, Drachman DA, Lippa CF. Apolipoprotein E epsilon 4 allele and the lifetime risk of Alzheimer's disease: what physicians know, and what they should know. Arch Neurol 1995; 52:1074–9.

25. Meyer MR, Tschanz JT, Norton MC, et al. APOE genotype predicts when—not whether—one is predisposed to develop Alzheimer disease. Nat Genet 1998; 19:321–2.

26. Farrer LA, Cupples LA, Haines JL, et al. Effects of age, sex, and ethnicity on the association between apolipoprotein E genotype and Alzheimer disease: a meta-analysis. JAMA 1997; 278:1349–56.

27. Myers AJ, Goate AM. The genetics of late-onset Alzheimer's disease. Curr Opin Neurol 2001; 14:433–40.

28. Saunders AM, Hulette O, Welsh-Bohmer KA, et al. Specificity, sensitivity, and predictive value of apolipoprotein-E genotyping for sporadic Alzheimer's disease. Lancet 1996; 348:90–3.

29. Chapman PF, Falinska AM, Knevett SG, Ramsay MF. Genes, models and Alzheimer's disease. Trends Genet 2001; 17:254–61.

30. Citron M. Beta-secretase as a target for the treatment of Alzheimer's disease. J Neurosci Res 2002; 70:373–9.

31. Moechars D, Lorent K, De Strooper B, Dewachter I, Van Leuven F. Expression in brain of amyloid precursor protein mutated in the α-secretase site causes disturbed behavior, neuronal degeneration and premature death in transgenic mice. EMBO J 1996; 15:1265–74.

32. Holcomb L, Gordon MN, McGowan E, et al. Accelerated Alzheimer-type phenotype in transgenic mice carrying both mutant amyloid precursor protein and presenilin 1 transgenes. Nat Med 1998; 4:97–100.

33. Schenk D, Barbour R, Dunn W, et al. Immunization with amyloid-beta attenuates Alzheimer-disease-like pathology in the PDAPP mouse. Nature 1999; 400:173–7.

34. Morgan D, Diamond DM, Gottschall PE, et al. A beta peptide vaccination prevents memory loss in an animal model of Alzheimer's disease. Nature 2000; 408:982–5. [Erratum, Nature 2001; 412:660.]

35. Helmuth L. Alzheimer's congress: further progress on a beta-amyloid vaccine. Science 2000; 289:375.

36. Check E. Nerve inflammation halts trial for Alzheimer's drug. Nature 2002; 415:462.

37. Birmingham K, Frantz S. Set back to Alzheimer vaccine studies. Nat Med 2002; 8:199–200.

38. De Strooper B, Konig G. An inflammatory drug prospect. Nature 2001; 414:159–60.

39. Hoehn MM, Yahr MD. Parkinsonism: onset, progression and mortality. Neurology 1967; 17: 427–42.

40. Pollanen MS, Dickson DW, Bergeron C. Pathology and biology of the Lewy body. J Neuropathol Exp Neurol 1993; 52:183–91.

41. Kuzuhara S, Mori H, Izumiyama N, Yoshimura M, Ihara Y. Lewy bodies are ubiquitinated: a light and electron microscopic immunocytochemical study. Acta Neuropathol (Berl) 1988; 75:345–53.

42. Tanner CM, Goldman SM. Epidemiology of Parkinson's disease. Neurol Clin 1996; 14:317–35.

43. Hughes AJ, Daniel SE, Kilford L, Lees AJ. Accuracy of clinical diagnosis of idiopathic Parkinson's disease: a clinico-pathological study of 100 cases. J Neurol Neurosurg Psychiatry 1992; 55:181–4.

44. Tanner CM, Ottman R, Goldman SM, et al. Parkinson disease in twins: an etiologic study. JAMA 1999; 281:341–6.

45. Duvoisin RC, Johnson WG. Hereditary Lewy-body parkinsonism and evidence for a genetic etiology of Parkinson's disease. Brain Pathol 1992; 2:309–20.

46. Burn DJ, Mark MH, Playford ED, et al. Parkinson's disease in twins studied with 18F-dopa and positron emission tomography. Neurology 1992; 42:1894–900.

47. Morrish PK, Rakshi JS, Bailey DL, Sawle GV, Brooks DJ. Measuring the rate of progression and estimating the preclinical period of Parkinson's disease with [18F]dopa PET. J Neurol Neurosurg Psychiatry 1998; 64:314–9.

48. Marder K, Tang M-X, Mejia H, et al. Risk of Parkinson's disease among first-degree relatives: a community-based study. Neurology 1996; 47:155–60.

49. Payami H, Larsen K, Bernard S, Nutt J. Increased risk of Parkinson's disease in parents and siblings of patients. Ann Neurol 1994; 36:659–61.

50. Piccini P, Morrish PK, Turjanski N, et al. Dopaminergic function in familial Parkinson's disease: a clinical and 18F-dopa positron emission tomography study. Ann Neurol 1997; 41:222–9.

51. Duvoisin RC. Recent advances in the genetics of Parkinson's disease. Adv Neurol 1996; 69:33–40.

52. Polymeropoulos MH, Lavedan C, Leroy E, et al. Mutation in the alpha-synuclein gene identified in families with Parkinson's disease. Science 1997; 276:2045–7.

53. Kruger R, Kuhn W, Muller T, et al. Ala30Pro mutation in the gene encoding alpha-synuclein in Parkinson's disease. Nat Genet 1998; 18:106–8.

54. Spillantini MG, Schmidt ML, Lee VM, Trojanowski JQ, Jakes R, Goedert M. α-Synuclein in Lewy bodies. Nature 1997; 388:839–40.

55. Mezey E, Dehejia AM, Harta G, et al. Alpha synuclein is present in Lewy bodies in sporadic Parkinson's disease. Mol Psychiatry 1998; 3:493–9. [Erratum, Mol Psychiatry 1999; 4:197.]

56. Conway KA, Lee SJ, Rochet JC, Ding TT, Williamson RE, Lansbury PT Jr. Acceleration of oligomerization, not fibrillization, is a shared property of both alpha-synuclein mutations linked to early-onset Parkinson's disease: implications for pathogenesis and therapy. Proc Natl Acad Sci U S A 2000; 97:571–6.

57. Louis ED, Fahn S. Pathologically diagnosed diffuse Lewy body disease and Parkinson's disease: do the parkinsonian features differ? Adv Neurol 1996; 69:311–4.

58. Kosaka K, Iseki E. Dementia with Lewy bodies. Curr Opin Neurol 1996; 9:271–5.

59. Spillantini MG, Crowther RA, Jakes R, Cairns NJ, Lantos PL, Goedert M. Filamentous alpha-synuclein inclusions link multiple system atrophy with Parkinson's disease and dementia with Lewy bodies. Neurosci Lett 1998; 251:205–8.

60. Galvin JE, Lee VM, Trojanowski JQ. Synucleinopathies: clinical and pathological implications. Arch Neurol 2001; 58:186–90.

61. Matsumine H, Saito M, Shimoda-Matsubayashi S, et al. Localization of a gene for an autosomal recessive form of juvenile parkinsonism to chromosome 6q25.2–27. Am J Hum Genet 1997; 60:588–96.

62. Kitada T, Asakawa S, Hattori N, et al. Mutations in the parkin gene cause autosomal recessive juvenile parkinsonism. Nature 1998; 392:605–8.

63. Abbas N, Lucking CB, Ricard S, et al. A wide variety of mutations in the parkin gene are responsible for autosomal recessive parkinsonism in Europe. Hum Mol Genet 1999; 8:567–74.

64. Lucking CB, Abbas N, Durr A, et al. Homozygous deletions in parkin gene in European and North African families with autosomal recessive juvenile parkinsonism. Lancet 1998; 352:1355–6.

65. Lücking CB, Dürr A, Bonifati V, et al. Association between early-onset Parkinson's disease and mutations in the *parkin* gene. N Engl J Med 2000; 342:1560–7.

66. Farrer M, Chan P, Chen R, et al. Lewy bodies and parkinsonism in families with parkin mutations. Ann Neurol 2001; 50:293–300.

67. Shimura H, Schlossmacher MG, Hattori N, et al. Ubiquitination of a new form of alpha-synuclein by parkin from human brain: implications for Parkinson's disease. Science 2001; 293:263–9.

68. Leroy E, Boyer R, Auburger G, et al. The ubiquitin pathway in Parkinson's disease. Nature 1998; 395:451–2.

69. Farrer M, Gwinn-Hardy K, Muenter M, et al. A chromosome 4p haplotype segregating with Parkinson's disease and postural tremor. Hum Mol Genet 1999; 8:81–5.

70. Gasser T, Muller-Myhsok B, Wszolek ZK, et al. A susceptibility locus for Parkinson's disease maps to chromosome 2p13. Nat Genet 1998; 18:262–5.

71. Funayama M, Hasegawa K, Kowa H, Saito M, Tsuji S, Obata F. A new locus for Parkinson's disease (PARK8) maps to chromosome 12p11.2-q13.1. Ann Neurol 2002; 51:296–301.

72. Valente EM, Bentivoglio AR, Dixon PH, et al. Localization of a novel locus for autosomal recessive

early-onset parkinsonism, PARK6, on human chromosome 1p35–p36. Am J Hum Genet 2001; 68:895–900.

73. van Duijn CM, Dekker MC, Bonifati V, et al. Park7, a novel locus for autosomal recessive early-onset parkinsonism, on chromosome 1p36. Am J Hum Genet 2001; 69:629–34.

74. Bonifati V, Rizzu P, van Baren MJ, et al. Mutations in DJ-1 gene associated with autosomal recessive early-onset parkinsonism. Science 2003; 299:256–9.

75. Scott WK, Nance MA, Watts RL, et al. Complete genomic screen in Parkinson disease: evidence for multiple genes. JAMA 2001; 286:2239–44.

76. Pankratz N, Nichols WC, Uniacke SK, et al. Genome screen to identify susceptibility genes for Parkinson disease in a sample without *parkin* mutations. Am J Hum Genet 2002; 71:124–35.

77. Hicks AA, Petursson H, Jonsson T, et al. A susceptibility gene for late-onset idiopathic Parkinson's disease. Ann Neurol 2001; 52:549–55.

78. Sveinbjörnsdóttir S, Hicks AA, Jónsson T, et al. Familial aggregation of Parkinson's disease in Iceland. N Engl J Med 2000; 343:1765–70.

79. Tan EK, Matsuura T, Nagamitsu S, Khajavi M, Jankovic J, Ashizawa T. Polymorphism of NACP-Rep1 in Parkinson's disease: an etiologic link with essential tremor? Neurology 2000; 54:1195–8.

80. Kruger R, Vieira-Saecker AM, Kuhn W, et al. Increased susceptibility to sporadic Parkinson's disease by a certain combined alpha-synuclein/apolipoprotein E genotype. Ann Neurol 1999; 45:611–7.

81. Farrer M, Maraganore DM, Lockhart P, et al. Alpha-synuclein gene haplotypes are associated with Parkinson's disease. Hum Mol Genet 2001; 10: 1847–51.

82. Izumi Y, Morino H, Oda M, et al. Genetic studies in Parkinson's disease with an alpha-synuclein/NACP gene polymorphism in Japan. Neurosci Lett 2001; 300:125–7.

10

Molecular Diagnosis of the Hematologic Cancers

LOUIS M. STAUDT, M.D., Ph.D.

The diagnosis of the hematologic cancers presents a daunting challenge. The many stages of normal hematopoietic differentiation give rise to a number of biologically and clinically distinct cancers. Inherited DNA-sequence variants do not appear to have a prominent causative role; rather, these diverse cancers are typically initiated by acquired alterations to the genome of the cancer cell, such as chromosomal translocations, mutations, and deletions. The diagnosis of the hematologic cancers is commonly based on morphologic evaluation supplemented by analysis of a few molecular markers. However, in some diagnostic categories defined in this fashion, the response of patients to treatment is markedly heterogeneous, arousing the suspicion that there can be several molecularly distinct diseases within the same morphologic category.

Gene-expression profiling is a genomics technique that has proved effective in deciphering this biologic and clinical diversity. The approach relies on the fact that only a fraction of the genes encoded in the genome of each cell are expressed—that is, actively transcribed into messenger RNA (mRNA) (Fig. 10.1A). The abundance of mRNA for each gene depends on a cell's lineage and stage of differentiation, on the activity of intracellular regulatory pathways, and on the influence of extracellular stimuli. To a large extent, the complement of mRNAs in a cell dictates its complement of proteins, and consequently, gene expression is a major determinant of the biology of normal and malignant cells.

In the process of expression profiling, robotically printed DNA microarrays are used to measure the expression of tens of thousands of genes at a time; this creates a molecular profile of the RNA in a tumor sample[1] (Fig. 10.1B). A variety of analytic techniques are used to classify cancers on the basis of their gene-expression profiles.[2,3] There are two

Originally published May 1, 2003

general approaches. In an unsupervised approach, pattern-recognition algorithms are used to identify subgroups of tumors that have related gene-expression profiles (Fig. 10.2A). In a supervised approach, statistical methods are used to relate gene-expression data and clinical data (Fig. 10.2B). These methods have revealed unexpected subgroups within the diagnostic categories of the hematologic cancers that are based on morphology and have demonstrated that the response to therapy is dictated by multiple independent biologic features of a tumor. This is not a comprehensive review of hematologic cancers; rather, it will provide examples of how gene-expression profiling has been used to provide a framework for the molecular diagnosis of these cancers.

MOLECULAR DIAGNOSIS OF NON-HODGKIN'S LYMPHOMA

Diffuse Large-B-Cell Lymphoma

Some cases of diffuse large-B-cell lymphoma respond well to multiagent chemotherapy,[5] but this lymphoma nonetheless remains a perplexing clinical puzzle, since roughly 60 percent of cases are incurable. This observation raises the possibility that this single diagnostic category may harbor more than one molecular disease.

The gene-expression profiles of lymph-node–biopsy specimens from patients with morphologically identical diffuse large-B-cell lymphoma show pronounced variability, with no common set of genes expressed in all cases.[4,6,7] To make sense of this variability, genes were classified into expression signatures[8]—that is, groups of genes with similar patterns of expression in a set of samples. Some signatures include genes expressed in a particular type of cell or stage of differentiation, whereas other signatures include genes expressed during a particular biologic response, such as cellular proliferation or the activation of a cellular signaling pathway.

One gene-expression signature that varies markedly among diffuse large-B-cell lymphomas is the germinal-center B-cell signature.[4,6] This signature characterizes B cells that are responding to a foreign antigen within the germinal-center microenvironment of secondary lymphoid organs. Among biopsy samples from patients with diffuse large-B-cell lymphoma, three biologically and clinically distinct subgroups have been identified[4,6] (Fig. 10.3A). The germinal-center B-cell–like subgroup (approximately 50 percent of cases) has high levels of expression of germinal-center B-cell signature genes, whereas the other two subgroups of diffuse large-B-cell lymphoma—termed activated B-cell–like and type 3—do not. The activated B-cell–like subgroup (approximately 30 percent of cases) instead expresses genes that are induced by mitogenic stimulation of blood B cells. The type 3 subgroup does not express genes characteristic of the other two subgroups and may yet be found to be heterogeneous. These findings suggest that the subgroups of diffuse large-B-cell lymphoma arise from different stages of normal B-cell development.

The notion that the gene-expression subgroups represent pathogenetically distinct types of diffuse large-B-cell lymphoma has been strongly supported by analysis of recurring chromosomal abnormalities in this cancer.[4,10] The t(14;18) translocation involving the *BCL2* gene and the amplification of the *c-rel* gene on chromosome 2p are recurrent oncogenic events in germinal-center B-cell–like diffuse large-B-cell lymphoma, but they

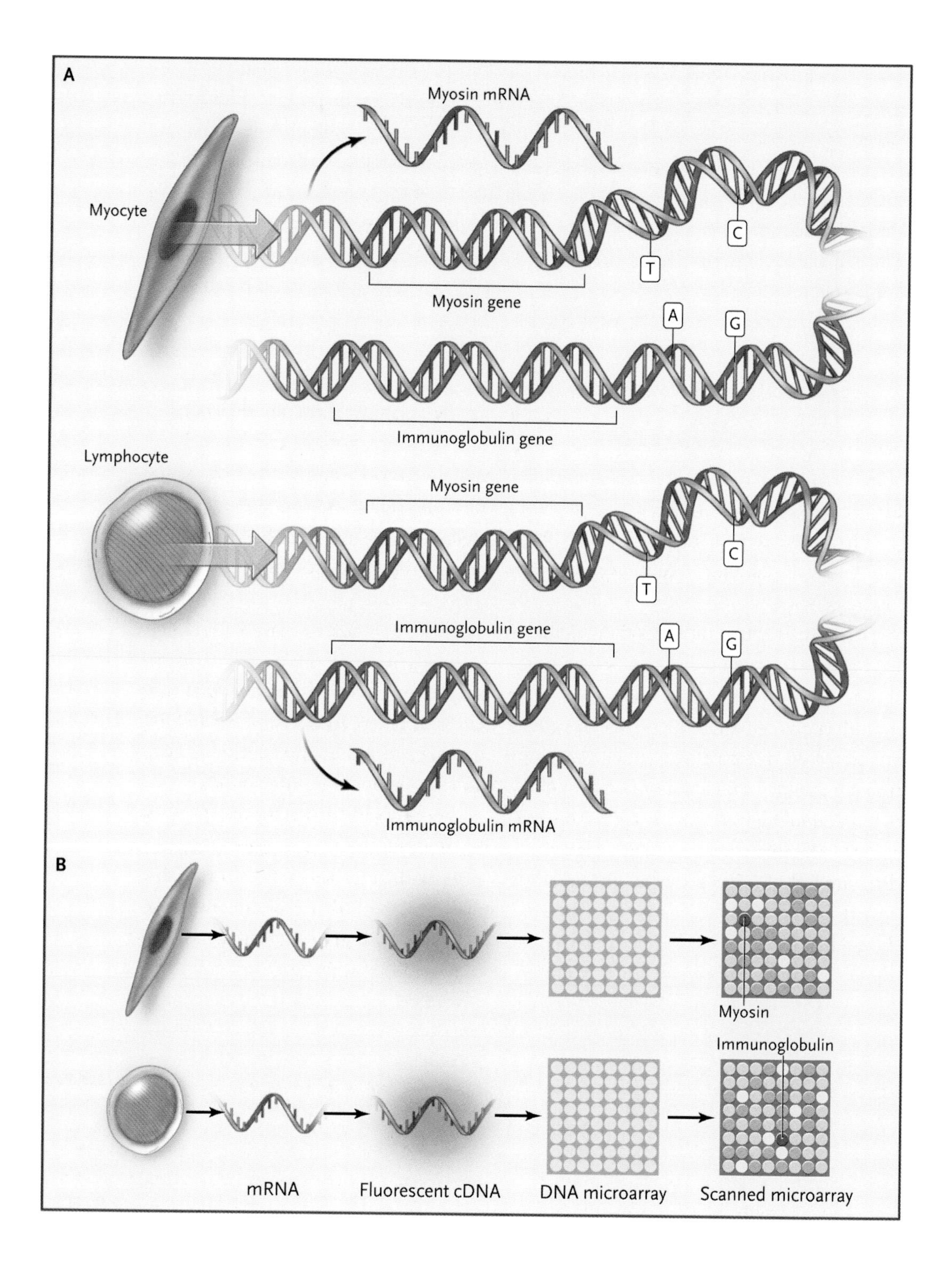
A
Myosin mRNA
Myocyte
T
C
Myosin gene
A
G
Immunoglobulin gene
Lymphocyte
Myosin gene
T
C
Immunoglobulin gene
A
G
Immunoglobulin mRNA
B
Myosin
Immunoglobulin
mRNA
Fluorescent cDNA
DNA microarray
Scanned microarray

never occur in the other subgroups. Activation of the nuclear factor-κB signaling pathway is a feature of the activated B-cell–like subgroup but not the other subgroups, and interference with this pathway selectively kills this type of diffuse large-B-cell lymphoma.[11]

The subgroups defined with the use of gene-expression signatures are clinically distinct as well: patients with the germinal-center B-cell–like form have a higher rate of overall survival five years after chemotherapy than do patients in the other subgroups[4,6] (Fig. 10.3A). This clinical distinction based on gene-expression profiles was evident even after the patients were classified according to the International Prognostic Index,[4,6] a well-established predictor of outcome in diffuse large-B-cell lymphoma.[12]

Predicting the Clinical Outcome

The example of diffuse large-B-cell lymphoma demonstrates how an unsupervised analysis of gene-expression data can reveal clinically distinct subgroups of tumors. In the complementary, supervised approach, clinical data are used to identify genes whose patterns of expression are correlated with the length of survival after diagnosis or with the likelihood that therapy will be curative. This approach has been used to develop robust predictors of prognosis in mantle-cell lymphoma[13] and diffuse large-B-cell lymphoma.[4,7]

Mantle-cell lymphoma constitutes approximately 8 percent of cases of non-Hodgkin's lymphomas but a much larger fraction of deaths from lymphoma, since current therapy is not curative. The length of survival among patients with mantle-cell lymphoma is quite variable, ranging from less than 1 year to more than 10 years.[13] Gene-expression profiling revealed a strong association between the expression of genes in the "proliferation" signature and survival in mantle-cell lymphoma.[13] The proliferation signature includes genes that are more highly expressed in dividing cells than in quiescent cells (Fig. 10.4A). The quartile of patients with the lowest level of proliferation-signature expression had a median survival of 6.7 years, whereas the quartile with the highest level of expression had a median survival of 0.8 year (Fig. 10.4A). The variable survival of patients with mantle cell lymphoma is therefore largely dictated by a single aspect of tumor biology, the rate of cell division, which can be quantitated by gene-expression profiling.

Figure 10.1 (facing page). Differential Expression of Messenger RNA (mRNA) by Different Types of Cells (Panel A) and Gene-Expression Profiling Using DNA Microarrays (Panel B). In Panel A, different types of cells, exemplified by a myocyte and a lymphocyte, express a distinct set of mRNAs from their genomes. Although the myocyte and lymphocyte possess the same inherited genomic DNA, distinct regulatory networks inside each cell cause different genes to be expressed as mRNA. The genes that encode myosin and immunoglobulin are among the most differentially expressed genes between these two types of cells. The mRNAs for other genes may be present in both types of cells, but at different levels, which may also affect the biology of the cells. Panel B shows the technique of gene-expression profiling, which uses DNA microarrays. First, mRNA is extracted from a cell and copied enzymatically to create a fluorescent complementary DNA (cDNA) probe representing the expressed genes in the cell. This probe is then incubated on the surface of a DNA microarray, which contains spots of DNA derived from thousands of distinct human genes. During the incubation, each cDNA molecule in the probe hybridizes to the microarray spot that represents its respective gene. The extent of hybridization of fluorescent cDNAs to each microarray spot is quantitated with use of a scanning fluorescence microscope. The levels of expression of more than 20,000 genes—in this example, the genes for myosin and immunoglobulin—can be measured in a single DNA-microarray experiment.

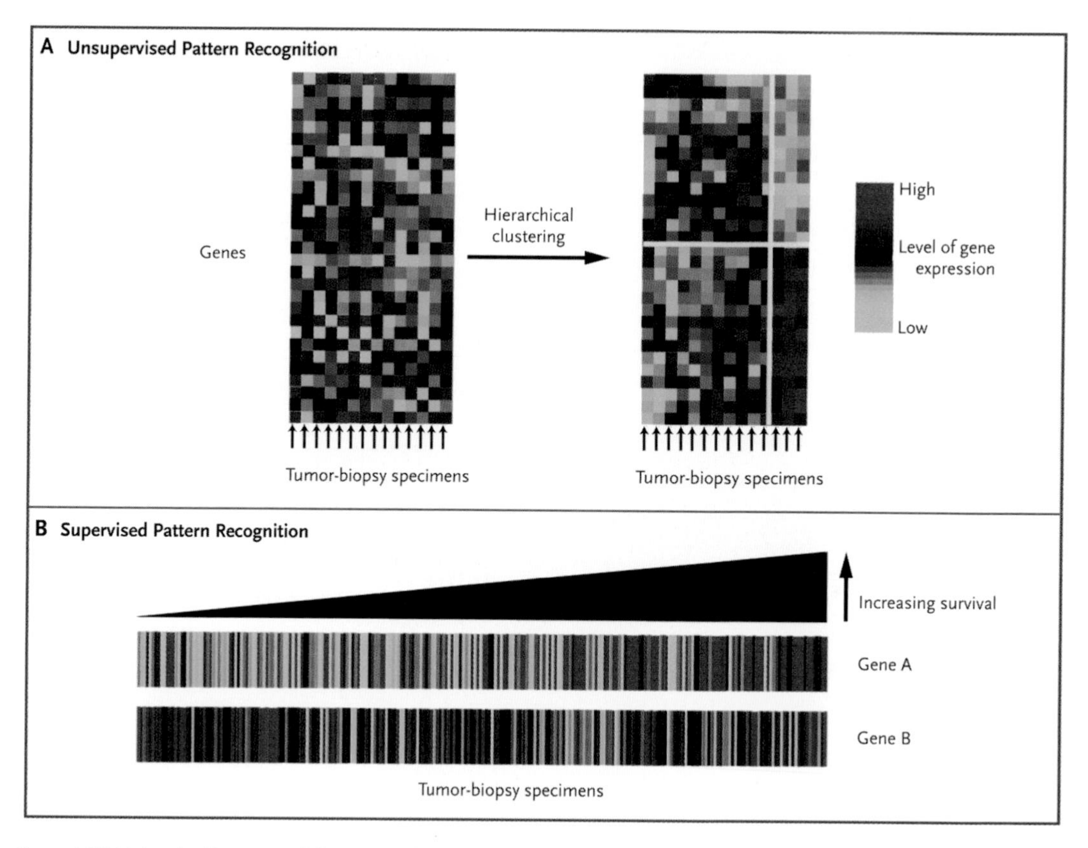

Figure 10.2 Molecular Diagnosis of Cancer by Gene-Expression Profiling with the Use of Unsupervised (Panel A) and Supervised (Panel B) Pattern-Recognition Algorithms. Panel A shows the discovery of cancer subgroups with the use of an unsupervised pattern-recognition algorithm. The expression of genes, as determined by DNA-microarray analysis, is depicted in a tabular format. Each row represents data for a particular human gene, and each column represents the expression of genes in a single biopsy sample (arrows). Highly expressed genes are shown in shades of red, and less highly expressed genes are shown in shades of green, according to the color scale shown. Before the analysis, no pattern is apparent (left-hand panel). A mathematical algorithm, termed "hierarchical clustering,"[2] is applied to the gene-expression data to search for a pattern (right-hand panel). This algorithm first rearranges the genes (in rows) so that genes with related patterns of expression are clustered. The algorithm next rearranges the samples (in columns) so that samples that have related expression of these genes are clustered. In this example, the hierarchical-clustering algorithm identified a clear subgroup of three tumor samples (on the far right-hand side) whose pattern of gene expression is distinct. Panel B shows how a supervised statistical algorithm is used to identify genes with patterns of expression that predict the clinical outcome. For each gene on the microarray, expression data from tumors are correlated with overall survival data from the corresponding patients. The example shows two genes with patterns of expression that are correlated with survival after chemotherapy for diffuse large-B-cell lymphoma.[4] A high level of expression of gene A is associated with extended survival, whereas a high level of expression of gene B is associated with short survival. Neither gene has a pattern of expression that is perfectly correlated with survival, illustrating that the clinical outcome is independently influenced by multiple molecular and clinical variables.[4]

Although the subgroups of diffuse large-B-cell lymphoma have distinct survival rates, the statistical approach of supervised analysis identified additional molecular differences among the tumors that can account for much of the remaining heterogeneity in survival[4,7] (Fig. 10.4B). This approach demonstrated that at least five distinct features of diffuse large-B-cell lymphomas influence the response to chemotherapy.[4] Specifically, the levels of expression of the germinal-center B-cell signature, the proliferation signature,

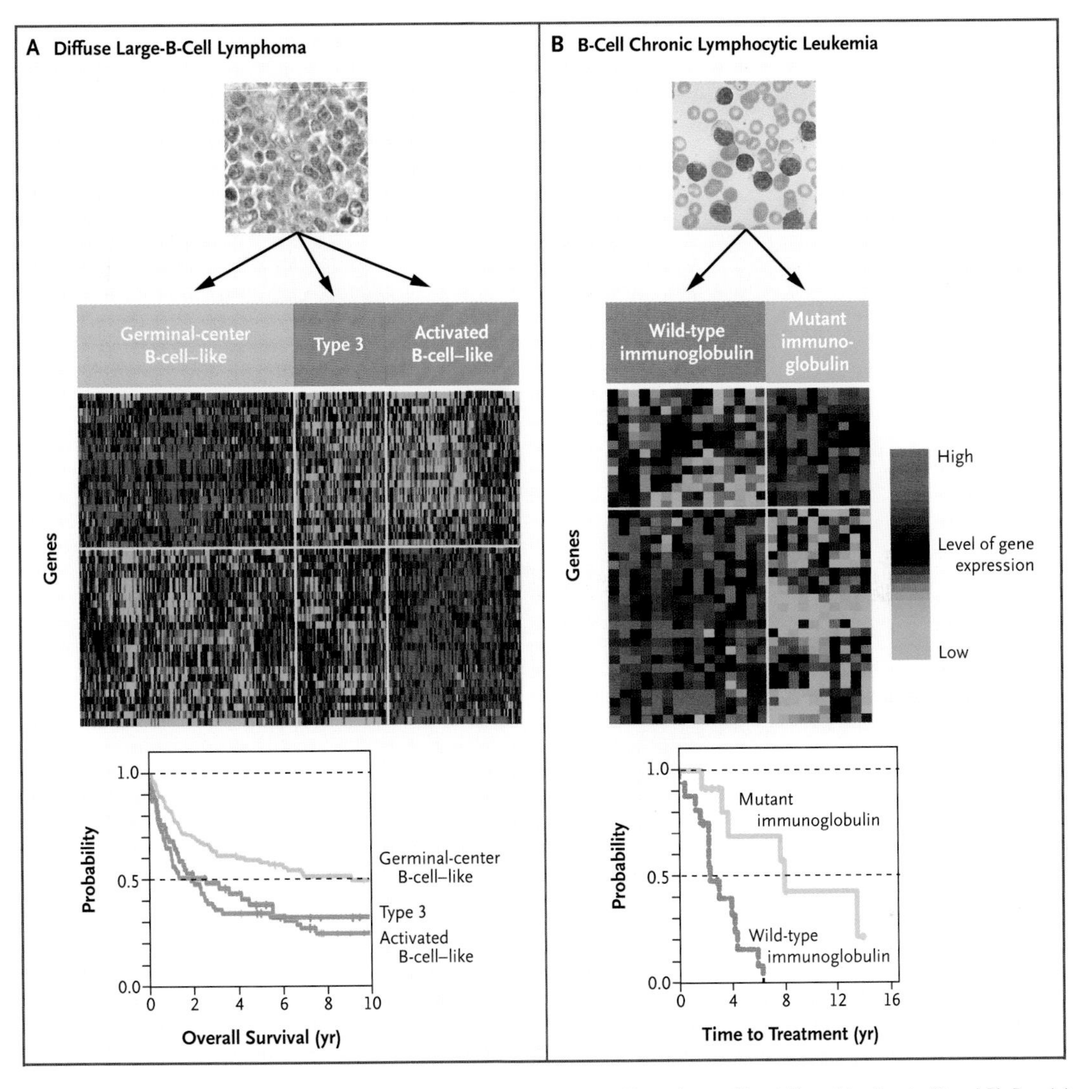

Figure 10.3 Examples of Molecularly and Clinically Distinct Subgroups of Lymphoma (Panel A) and Leukemia (Panel B). Panel A shows the levels of expression of 57 genes that distinguish three subgroups of diffuse large-B-cell lymphoma[4]: germinal-center B-cell–like (orange), type 3 (purple), and activated B-cell–like (blue). The Kaplan–Meier curve shows that overall survival differs among the subgroups after chemotherapy. Panel B shows 39 genes that are differentially expressed in two subgroups of B-cell chronic lymphocytic leukemia,[9] one with unmutated (wild-type) immunoglobulin genes (purple) and one with somatically mutated immunoglobulin genes (blue). The Kaplan–Meier curve shows that the two subgroups differ with respect to the time to initial treatment after diagnosis.

the major-histocompatibility-complex (MHC) class II signature, and the lymph-node signature were predictive of the clinical outcome, as was the level of expression of *BMP6*, a gene that does not belong to a defined expression signature. As in mantle-cell lymphoma, expression of the proliferation signature predicted a poor outcome. Predictive genes in two other signatures suggest that the host immune response has an important role in curative responses to chemotherapy. Expression of the lymph-node–signature genes reflects the nontumor cells in the diffuse large-B-cell lymphoma–biopsy specimen, including activated macrophages, natural killer cells, and stromal cells. A high

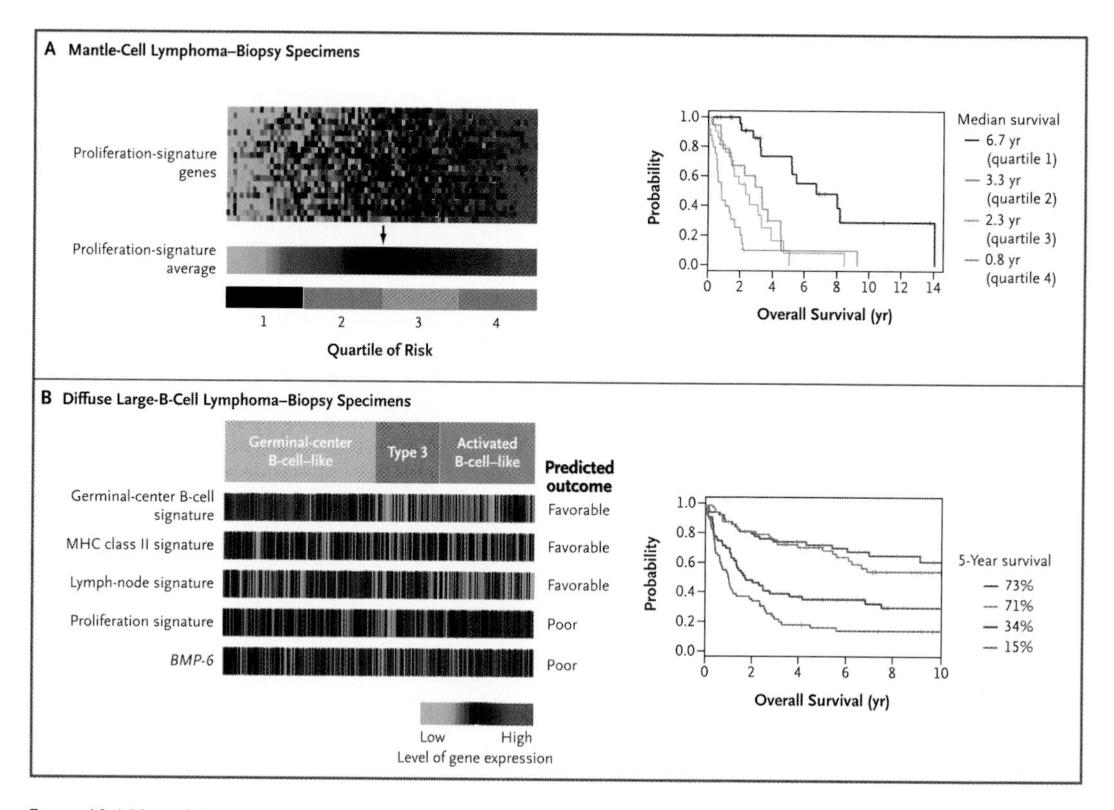

Figure 10.4 Use of the Proliferation Gene-Expression Signature to Predict the Clinical Outcome in Mantle-Cell Lymphoma (Panel A) and the Development of a Multivariate Gene-Expression–Based Predictor of Survival after Chemotherapy for Diffuse Large-B-Cell Lymphoma (Panel B). Panel A shows the use of the proliferation gene-expression signature to predict the length of survival in patients with mantle-cell lymphoma. Elevated levels of expression of genes in the proliferation gene-expression signature in a biopsy specimen of mantle-cell lymphoma was associated with short survival.[13] The relative level of expression of the proliferation-signature genes is represented by the color bars; the biopsy samples are ordered from left to right according to the increasing relative expression of the proliferation-signature genes. The levels of expression of 20 proliferation-signature genes were averaged, and the resulting average was used to subdivide patients with mantle-cell lymphoma into four quartiles. The Kaplan–Meier plot illustrates the striking differences in the length of survival among these four risk groups. In Panel B, the biopsy specimens of diffuse large-B-cell lymphoma are ordered as in Figure 10.3A according to their assignment to the three subgroups. A supervised analysis of gene-expression data identified four gene-expression signatures and one single gene—*BMP6*—with patterns of expression that correlated with clinical outcome.[4] A high level of expression of a gene or signature within a tumor was associated with a favorable or poor outcome after chemotherapy, as indicated. The colored bars represent the relative levels of expression of each signature or gene in each of the biopsy specimens according to the scale shown. The levels of expression of the signatures represent averages of data from multiple genes in each signature. These five patterns of gene expression vary independently of one another. Since each of these patterns correlates with the clinical outcome, multiple biologic attributes of the tumors must influence the clinical outcome. A linear combination of these five gene-expression components is used to assign a gene-expression outcome-predictor score for each patient. Patients are ranked according to their outcome-predictor scores and divided into quartiles. The Kaplan–Meier plot demonstrates the ability of the gene-expression–based outcome predictor to classify patients with diffuse large-B-cell lymphoma into prognostic groups. MHC denotes major histocompatibility complex. Data are adapted from Lymphoma/Leukemia Molecular Profiling Project studies of gene expression and clinical outcome in patients with diffuse large-B-cell lymphoma and mantle-cell lymphoma.[4,13]

level of expression of these genes predicts a favorable clinical outcome, suggesting that this reactive immune response is beneficial. The MHC class II signature includes genes encoding components of this critical antigen-presentation–protein complex, and decreased expression of these genes predicts a poor outcome. These findings suggest that some tumors may evade the immune response by down-regulating their antigen-presentation capacity.

These expression signatures can be combined to form a multivariate predictor of survival after chemotherapy for diffuse large-B-cell lymphoma.[4] With the use of this approach, half the patients can be placed into a favorable-risk group, with a five-year survival rate of more than 70 percent; one quarter can be assigned to a poor-risk group, with a five-year survival rate of 15 percent; and the remaining patients are in an intermediate-risk group, with a five-year survival rate of 34 percent (Fig. 10.4B).

MOLECULAR DIAGNOSIS OF LEUKEMIAS

Acute Leukemias

The molecular diagnosis of leukemias began with the recognition and analysis of recurrent chromosomal translocations.[14,15] The genes discovered at the translocation break points have drawn attention to critical regulatory pathways in hematopoietic cells that can cause cancer when they are dysregulated. In many acute leukemias, translocations fuse genes that reside on the two partner chromosomes, creating a chimeric gene with novel oncogenic properties.

Chromosomal translocations have been used to identify patients with acute leukemia with distinct clinical outcomes.[16,17] In acute myeloid leukemia (AML), for instance, the presence of a t(8;21) translocation or a chromosome 16 inversion identifies patients with a comparatively good prognosis, whereas the t(9;22) translocation is associated with a poor outcome.[17] It is important to note that chromosomal translocations have been used to identify patients who will benefit from intensifying the dose of chemotherapy.[18–20]

Despite these prognostic and therapeutic successes, chromosomal translocations account for only part of the varied clinical behavior of acute leukemia, for several reasons. First, other genetic aberrations can be functionally equivalent to a translocation,[21,22] thus diminishing the prognostic power of a translocation as a single variable. Second, additional oncogenic abnormalities may accumulate in a leukemia that alter its responsiveness to therapy. For example, mutations in the gene encoding the flt3 receptor tyrosine kinase have been associated with response to treatment in patients with AML.[23–26] Furthermore, flt3 mutations that activate the kinase are present in some cases of acute lymphoblastic leukemia (ALL) with a t(4;14) translocation, rendering them susceptible to killing by flt3 inhibitors.[27] Finally, a sizable fraction of the acute leukemias have none of the defined recurrent translocations.[16,17]

Gene-expression profiling has been used as an alternative approach to mapping chromosomal translocations. In pediatric B-cell ALL, gene-expression signatures have been identified that correlate with six different chromosomal abnormalities.[28,29] These gene-expression signatures can be combined with the use of statistical algorithms to predict chromosomal abnormalities with 96 to 100 percent accuracy.[29] Likewise, in adult AML, a gene-expression–based predictor has been created that can identify three different chromosomal translocations with a high rate of accuracy.[30] Gene-expression predictors can also identify patients with AML who have isolated trisomy 8.[31] These encouraging results demonstrate that DNA microarrays can be used to diagnose most chromosomal abnormalities in acute leukemias and could potentially substitute for the multiple diagnostic tests for these abnormalities that are currently required.

An oncogene likely to be causally related to T-cell ALL can be dysregulated by chromosomal translocations in some cases but by alternative mechanisms in others.[22] For example, the *HOX11* oncogene is involved in recurrent but infrequent translocations in T-cell ALL, but gene-expression profiling revealed that some cases of T-cell ALL overexpress *HOX11* without any detectable chromosomal abnormalities in this gene. All leukemias that overexpress *HOX11* have a common gene-expression signature, suggesting that they are biologically similar. Most important, patients with leukemias that overexpress *HOX11* have a favorable outcome, as compared with patients with other types of T-cell ALL, whether or not the overexpression is due to translocation, indicating the clinical superiority of expression profiling[22] over identification of the translocation.

Two adverse events after the treatment of acute leukemias are relapse and the development of secondary leukemias. In B-cell ALL, gene-expression profiling at the time of diagnosis provided information that could predict which patients would relapse and which would remain in continuous complete remission.[29] Interestingly, no patterns of gene expression have been found to predict relapse in all subtypes of ALL. Rather, relapse was predicted by the expression of different genes in each leukemic subtype, emphasizing once again their divergent biologic characteristics. Secondary AML arises as a consequence of treatment in some patients with ALL, and this complication could also be predicted on the basis of gene-expression profiling in the subgroup of B-cell ALL with the t(12;21) translocation.[29] Although these predictors of clinical outcome will need to be validated in independent data sets, these findings suggest that treatment stratification based on gene-expression profiling can be initiated at the time of the initial diagnosis of ALL.

Chronic Lymphocytic Leukemia

The most common leukemia in humans—chronic lymphocytic leukemia (CLL)—is an indolent but inexorable disease with no cure. Studies of immunoglobulin gene mutations in CLL cells raised the intriguing hypothesis that CLL might be two distinct diseases.[32,33] The presence of somatic mutations in the immunoglobulin genes of CLL cells defined a group of patients who had stable or slowly progressing disease requiring late or no treatment. By contrast, the absence of immunoglobulin gene mutations in CLL cells defined a group of patients who had a progressive clinical course requiring early treatment. These two subtypes of CLL may also differ with respect to oncogenic mechanisms, since deletion of the ATM locus on chromosome 11q is associated with the absence of immunoglobulin gene mutations in CLL[34–36] and with shortened survival in some patients.[37]

Despite these clinical and molecular differences between the subtypes of CLL, gene-expression profiling revealed that CLL cells express a common gene-expression signature that differentiates this form of leukemia from other lymphoid cancers and from normal lymphoid subpopulations.[9,38] This signature is shared by all cases of CLL, irrespective of the immunoglobulin gene mutation status, suggesting that CLL should be considered a single disease entity.

Nonetheless, given the clear clinical differences between the two subtypes of CLL, a hunt was made for genes that correlated with this distinction.[9,38] Roughly 160 genes were found whose levels of expression differed significantly between the two subtypes[9] (Fig. 10.3B). Expression of the single most discriminating gene, *ZAP-70*, distinguished these two subtypes with 93 percent accuracy.[9,39] Whereas analysis of the immunoglobulin gene sequence would be a challenging and expensive test to introduce into routine clinical practice, a quantitative reverse-transcriptase–polymerase-chain-reaction assay or protein-based assay for the expression of *ZAP-70* is feasible.[39,40]

TRANSLATING MOLECULAR DIAGNOSIS INTO A CLINICAL REALITY

What form of technology will be used for the molecular diagnosis of cancer in the future? Our experience with gene-expression profiling has taught us two clear lessons: multiple genes need to be studied to distinguish most types of cancer, and quantitative measurement of molecular differences among tumors results in clinically important diagnostic and prognostic distinctions. An important goal will therefore be to develop a platform for routine clinical diagnosis that can quantitatively measure the expression of a few hundred genes. Such a diagnostic platform would allow us quickly to translate what we have learned about important molecular subgroups within each hematologic cancer. As we design new clinical trials, however, we must include genomic-scale gene-expression profiling in order to identify the genes that influence the response to the agents under investigation. In this fashion, we can iteratively refine the molecular diagnosis of the hematologic cancers on the basis of new advances in treatment and thus eventually reach the goal of tailored therapies for molecularly defined diseases.

Acknowledgments

Supported by intramural research funds from the National Cancer Institute.

I am indebted to my colleagues in the Lymphoma/Leukemia Molecular Profiling Project for their collaboration and for stimulating discussions regarding molecular diagnosis in hematologic cancers.

References

1. Staudt LM, Brown PO. Genomic views of the immune system. Annu Rev Immunol 2000; 18:829–59.
2. Eisen MB, Spellman PT, Brown PO, Botstein D. Cluster analysis and display of genome-wide expression patterns. Proc Natl Acad Sci U S A 1998; 95:14863–8.
3. Golub TR, Slonim DK, Tamayo P, et al. Molecular classification of cancer: class discovery and class prediction by gene expression monitoring. Science 1999; 286:531–7.
4. Rosenwald A, Wright G, Chan WC, et al. The use of molecular profiling to predict survival after chemotherapy for diffuse large-B-cell lymphoma. N Engl J Med 2002; 346:1937–47.
5. DeVita VT Jr, Canellos GP, Chabner B, Schein P, Hubbard SP, Young RC. Advanced diffuse histiocytic lymphoma, a potentially curable disease. Lancet 1975; 1:248–50.
6. Alizadeh AA, Eisen MB, Davis RE, et al. Distinct types of diffuse large B-cell lymphoma identified by gene expression profiling. Nature 2000; 403:503–11.
7. Shipp MA, Ross KN, Tamayo P, et al. Diffuse large B-cell lymphoma outcome prediction by gene-expression profiling and supervised machine learning. Nat Med 2002; 8:68–74.

8. Shaffer AL, Rosenwald A, Hurt EM, et al. Signatures of the immune response. Immunity 2001; 15:375–85.

9. Rosenwald A, Alizadeh AA, Widhopf G, et al. Relation of gene expression phenotype to immunoglobulin mutation genotype in B cell chronic lymphocytic leukemia. J Exp Med 2001; 194:1639–47.

10. Huang JZ, Sanger WG, Greiner TC, et al. The t(14; 18) defines a unique subset of diffuse large B-cell lymphoma with a germinal center B-cell gene expression profile. Blood 2002; 99:2285–90.

11. Davis RE, Brown KD, Siebenlist U, Staudt LM. Constitutive nuclear factor kappaB activity is required for survival of activated B cell-like diffuse large B cell lymphoma cells. J Exp Med 2001; 194: 1861–74.

12. The International Non-Hodgkin's Lymphoma Prognostic Factors Project. A predictive model for aggressive non-Hodgkin's lymphoma. N Engl J Med 1993; 329:987–94.

13. Rosenwald A, Wright G, Wiestner A, et al. The proliferation gene expression signature is a quantitative integrator of oncogenic events that predicts survival in mantle cell lymphoma. Cancer Cell 2003; 3:185–97.

14. Rowley JD. The critical role of chromosome translocations in human leukemias. Annu Rev Genet 1998; 32:495–519.

15. Nowell PC. Progress with chronic myelogenous leukemia: a personal perspective over four decades. Annu Rev Med 2002; 53:1–13.

16. Ferrando AA, Look AT. Clinical implications of recurring chromosomal and associated molecular abnormalities in acute lymphoblastic leukemia. Semin Hematol 2000; 37:381–95.

17. Mrozek K, Heinonen K, Bloomfield CD. Clinical importance of cytogenetics in acute myeloid leukaemia. Best Pract Res Clin Haematol 2001; 14:19–47.

18. Bloomfield CD, Lawrence D, Byrd JC, et al. Frequency of prolonged remission duration after high-dose cytarabine intensification in acute myeloid leukemia varies by cytogenetic subtype. Cancer Res 1998; 58:4173–9.

19. Ayigad S, Kuperstein G, Zilberstein J, et al. TEL-AML1 fusion transcript designates a favorable outcome with an intensified protocol in childhood acute lymphoblastic leukemia. Leukemia 1999; 13:481–3.

20. Maloney K, McGavran L, Murphy J, et al. TEL-AML1 fusion identifies a subset of children with standard risk acute lymphoblastic leukemia who have an excellent prognosis when treated with therapy that includes a single delayed intensification. Leukemia 1999; 13:1708–12.

21. Pabst T, Mueller BU, Zhang P, et al. Dominant-negative mutations of CEBPA, encoding CCAAT/enhancer binding protein-alpha (C/EBPalpha), in acute myeloid leukemia. Nat Genet 2001; 27:263–70.

22. Ferrando AA, Neuberg DS, Staunton J, et al. Gene expression signatures define novel oncogenic pathways in T cell acute lymphoblastic leukemia. Cancer Cell 2002; 1:75–87.

23. Nakao M, Yokota S, Iwai T, et al. Internal tandem duplication of the flt3 gene found in acute myeloid leukemia. Leukemia 1996; 10:1911–8.

24. Hayakawa F, Towatari M, Kiyoi H, et al. Tandem-duplicated Flt3 constitutively activates STAT5 and MAP kinase and introduces autonomous cell growth in IL-3-dependent cell lines. Oncogene 2000; 19:624–31.

25. Yamamoto Y, Kiyoi H, Nakano Y, et al. Activating mutation of D835 within the activation loop of FLT3 in human hematologic malignancies. Blood 2001; 97:2434–9.

26. Gilliland DG, Griffin JD. Role of FLT3 in leukemia. Curr Opin Hematol 2002; 9:274–81.

27. Armstrong SA, Kung AL, Mabon ME, et al. Inhibition of FLT3 in MLL: validation of a therapeutic target identified by gene expression based classification. Cancer Cell 2003; 3:173–83.

28. Armstrong SA, Staunton JE, Silverman LB, et al. MLL translocations specify a distinct gene expression profile that distinguishes a unique leukemia. Nat Genet 2001; 30:41–7.

29. Yeoh E-J, Ross ME, Shurtleff SA, et al. Classification, subtype discovery, and prediction of outcome in pediatric acute lymphoblastic leukemia by gene expression profiling. Cancer Cell 2002; 1:133–43.

30. Schoch C, Kohlmann A, Schnittger S, et al. Acute myeloid leukemias with reciprocal rearrangements can be distinguished by specific gene expression profiles. Proc Natl Acad Sci U S A 2002; 99:10008–13.

31. Virtaneva K, Wright FA, Tanner SM, et al. Expression profiling reveals fundamental biological differences in acute myeloid leukemia with isolated trisomy 8 and normal cytogenetics. Proc Natl Acad Sci U S A 2001; 98:1124–9.

32. Damle RN, Wasil T, Fais F, et al. Ig V gene mutation status and CD38 expression as novel prognostic indicators in chronic lymphocytic leukemia. Blood 1999; 94:1840–7.

33. Hamblin TJ, Davis Z, Gardiner A, Oscier DG, Stevenson FK. Unmutated Ig V(H) genes are associated with a more aggressive form of chronic lymphocytic leukemia. Blood 1999; 94:1848–54.

34. Stankovic T, Stewart GS, Fegan C, et al. Ataxia telangiectasia mutated-deficient B-cell chronic lymphocytic leukemia occurs in pregerminal center cells and results in defective damage response and unrepaired chromosome damage. Blood 2002; 99:300–9.

35. Krober A, Seiler T, Benner A, et al. V(H) mutation status, CD38 expression level, genomic aberrations, and survival in chronic lymphocytic leukemia. Blood 2002; 100:1410–6.

36. Oscier DG, Gardiner AC, Mould SJ, et al. Multivariate analysis of prognostic factors in CLL: clinical stage, IGVH gene mutational status, and loss or mutation of the p53 gene are independent prognostic factors. Blood 2002; 100:1177–84.

37. Dohner H, Stilgenbauer S, Benner A, et al. Genomic aberrations and survival in chronic lymphocytic leukemia. N Engl J Med 2000; 343:1910–6.

38. Klein U, Tu Y, Stolovitzky GA, et al. Gene expression profiling of B cell chronic lymphocytic leukemia reveals a homogeneous phenotype related to memory B cells. J Exp Med 2001; 194:1625–38.

39. Wiestner A, Rosenwald A, Barry TS, et al. ZAP-70 expression identifies a chronic lymphocytic leukemia subtype with unmutated immunoglobulin genes, inferior clinical outcome, and distinct gene expression profile. Blood 2003; 101:4944–51.

40. Crespo M, Bosch F, Villamor N, et al. ZAP–70 expression as a surrogate for immunoglobulin-variable-region mutations in chronic lymphocytic leukemia. N Engl J Med 2003; 348:1764–75.

11

Breast and Ovarian Cancer

RICHARD WOOSTER, Ph.D., and BARBARA L. WEBER, M.D.

Despite years of intensive study and substantial progress in understanding susceptibility to breast and ovarian cancer, these diseases remain important causes of death in women. However, several recent critical advances—sequencing of the human genome and the development of high-throughput techniques for identifying DNA-sequence variants, changes in copy numbers, and global expression profiles—have dramatically accelerated the pace of research aimed at preventing and curing these diseases. We review some of the important discoveries in the genetics of breast and ovarian cancer, ongoing studies to isolate additional susceptibility genes, and early work on molecular profiling involving microarrays.

SUSCEPTIBILITY TO BREAST AND OVARIAN CANCER

In the United States, 10 to 20 percent of patients with breast cancer and patients with ovarian cancer have a first- or second-degree relative with one of these diseases.[1] Two major genes associated with susceptibility to breast and ovarian cancer—breast cancer susceptibility gene 1 (*BRCA1*) and breast cancer susceptibility gene 2 (*BRCA2*)—have been identified to date.[2,3] Mutations in either of these genes confer a lifetime risk of breast cancer of between 60 and 85 percent and a lifetime risk of ovarian cancer of between 15 and 40 percent.[4,5] However, mutations in these genes account for only 2 to 3 percent of all breast cancers,[6,7] and susceptibility alleles in other genes, such as *TP53*, *PTEN*, and *STK11/LKB1*, are even less common causes of breast and ovarian cancer (Fig. 11.1).

The prediction that there are common DNA-sequence variants that confer a small but appreciable enhanced risk of cancer has been validated with the recent discovery of the 1100delC mutation in the cell-cycle–checkpoint kinase gene (*CHEK2*).[9] This mutation was found in 1.1 percent of women without breast cancer, 1.4 percent of women with a

Originally published June 5, 2003

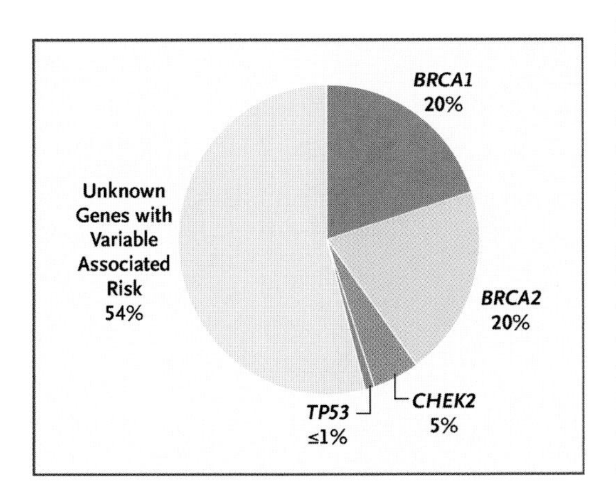

Figure 11.1 The Genetics of Breast Cancer. *BRCA1* and *BRCA2* mutations occur in approximately 20 percent of families with evidence of inherited susceptibility to breast cancer. Germ-line mutations in *TP53* cause the Li–Fraumeni syndrome and account for no more than 1 percent of cases of familial breast cancer, but women who survive the childhood cancers associated with the Li–Fraumeni syndrome have as much as a 90 percent risk of breast cancer.[8] Mutations in the cell-cycle–checkpoint kinase gene (*CHEK2*) account for about 5 percent of all cases of familial breast cancer (defined by the diagnosis of breast cancer in two or more family members before the age of 60 years), but the risk for individual mutation carriers is probably less than 20 percent.[9] All other cases of breast cancer are presumed to be due to an undefined number of additional susceptibility genes with various degrees of penetrance, exposure to hormonal and environmental factors, and stochastic genetic events.

personal but no family history of breast cancer, and 4.2 percent of index patients from 718 families in which two or more members had been given a diagnosis of breast cancer before the age of 60 years but in which there was no detectable *BRCA1* or *BRCA2* mutation. This mutation doubles the risk of breast cancer among women and increases the risk among men by a factor of 10. *CHEK2*, an important component of the cellular machinery that recognizes and repairs damaged DNA, is activated after phosphorylation by the checkpoint gene *ATM* and in turn activates BRCA1. The role of *ATM* mutations in the predisposition to the early onset of breast cancer remains controversial, but some missense mutations do appear to increase susceptibility to breast cancer in humans[10] and mice.[11]

There is convincing evidence that additional high-penetrance genes that increase susceptibility to breast cancer exist. In contrast, it has been suggested that, other than *BRCA1* and *BRCA2*, high-penetrance genes that confer susceptibility to ovarian cancer do not exist.[12] An ovarian-cancer–susceptibility locus on chromosome 3p22–25 has putatively been identified, but this finding has yet to be confirmed by an independent group.[13]

Many additional genetic variants in low-penetrance susceptibility alleles may moderately increase the risk of breast cancer, ovarian cancer, or both. These genetic variants are much more common in the population than are high-penetrance gene mutations and, thus, in aggregate may make a substantially greater contribution to breast and ovarian cancer in the population than mutations in high-risk genes.[14] However, genetic heterogeneity and the rarity of high-penetrance genes make both high- and low-penetrance genes difficult to identify.

IDENTIFICATION OF GENES THAT INCREASE SUSCEPTIBILITY TO BREAST AND OVARIAN CANCER

Identification of High-Penetrance Genes

Genetic linkage was used to identify the *BRCA1* and *BRCA2* loci on chromosomes 17q and 13q, respectively.[15,16] In both cases, no information was initially available on the location, structure, or function of the genes, and they were identified through positional cloning. Loss-of-heterozygosity mapping was of little assistance in either search; however, a homozygous deletion on chromosome 13 in a pancreatic adenocarcinoma helped identify the location of *BRCA2*.[17] Finally, critical data in the search for *BRCA2* came from studies of breast cancer in Iceland, whose population derives from a small group of settlers from Norway and Ireland.[18,19] Such populations share more genetic information than large, admixed populations and have been used successfully many times in gene mapping.[20] After the *BRCA1* locus was identified, it took almost four years to isolate the gene and involved several labor-intensive strategies.[2] By contrast, the *BRCA2* locus was one of the first genomic intervals to be systematically sequenced as part of the Human Genome Project. These data, together with other information about the genes in the region, reduced the time it took to isolate *BRCA2* to two years.[3] Thus, information from the human genome sequence greatly enhances the utility of linkage analysis for gene identification (Fig. 11.2).

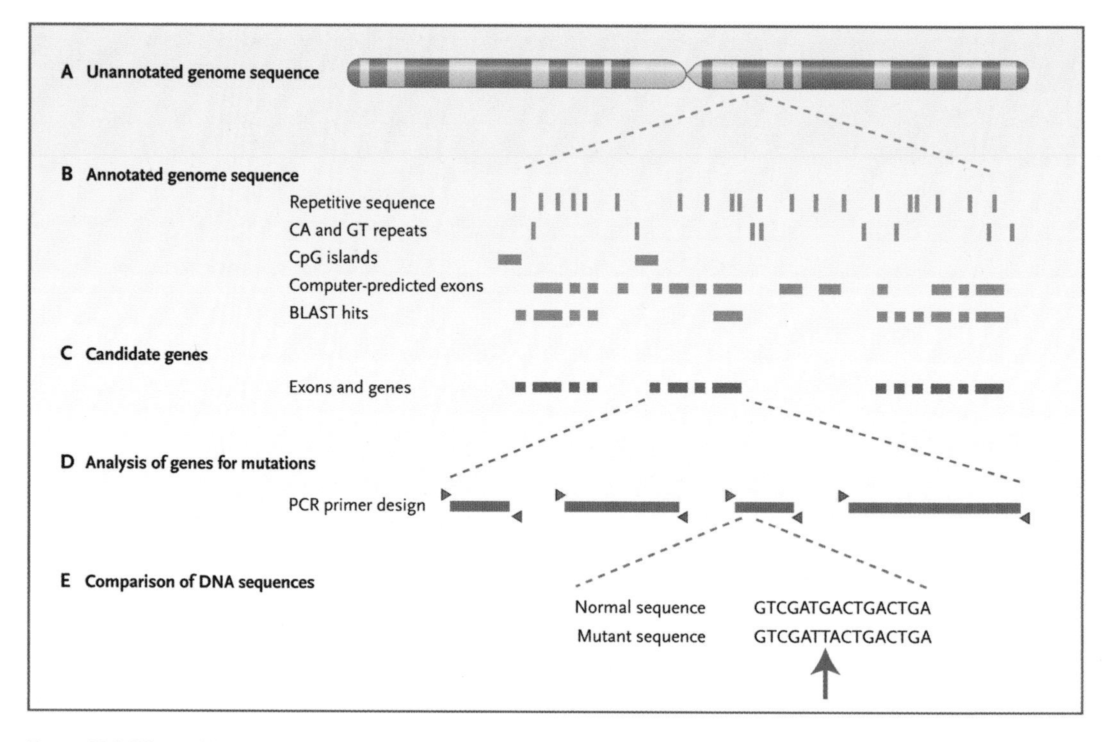

Figure 11.2 Effect of Sequencing the Human Genome on Gene-Discovery Strategies. The annotated DNA sequence of the human genome can be used to locate genes, repeat sequences, and other features and has revolutionized the identification of cancer genes. A sequence without annotation is of limited utility (Panel A). As shown in Panel B, an annotated sequence shows genetic markers such as CA and GT repeats along with other data, such as CpG islands, known genes, genes predicted to exist on the basis of computational models, and Basic Local Alignment Search Tool (BLAST) matches. Using publicly available data (http://www.ensembl.org, http://www.ncbi.nlm.nih.gov, and http://www.genome.ucsc.edu), it is possible to jump from a genetic region of interest to the identification of candidate genes in a matter of seconds and download the relevant data (Panel C). With these data in hand, experiments, such as those involving the polymerase chain reaction (PCR), can be designed to analyze the genes for mutations (Panel D). The final step in the identification of genes is to compare the sequence from patients with the disease of interest with the normal reference sequence to discover the mutations (Panel E).

BRCA3 and Beyond

Several candidate regions for *BRCA3* have been proposed, including chromosome 13q21[21] and chromosome 8p12–22,[22] but both have been strongly refuted by analysis of data from independent families.[23,24] The search for *BRCA3* has been difficult for several reasons. First, ovarian cancer and male breast cancer were recognized as components of syndromes of breast-cancer susceptibility before either *BRCA1* or *BRCA2* was isolated, allowing targeted identification of affected families. Since no such phenotype has been associated with the putative *BRCA3* gene or genes, families in current studies are selected only on the basis of a young age at the diagnosis of breast cancer and the absence of ovarian and male breast cancer. Ideally, these families should have multiple members with early-onset breast cancer and strong evidence against the involvement of either *BRCA1* or *BRCA2*. However, the breast cancers in most such families are in fact due to germ-line mutations in *BRCA1* or *BRCA2*,[25] and those that are not may represent the effects of multiple susceptibility alleles (genetic heterogeneity), reducing the power of linkage analysis. What is needed to further this effort are larger families to increase the statistical power of such studies, as well as novel means of clustering families into subgroups most likely to represent single-gene disorders.

One approach is to classify families with breast cancer according to the molecular profile of the associated tumors. These analyses could be based either on expression profiling or on array-based comparative genomic hybridization, both of which provide unique molecular signatures (Fig. 11.3). Nonetheless, hundreds of small pedigrees may be needed to identify the *BRCA3* locus.

Use of the Human Genome Sequence to Identify Low-Penetrance Genes

As noted, the susceptibility genes identified to date are not responsible for most breast and ovarian cancers, leaving a considerable potential contribution from less penetrant genes. One of the implicit problems in isolating low-penetrance genes is that such genes will rarely produce striking familial patterns involving multiple cases that can be used in traditional linkage studies. An additional concern is that very large studies, with statistical power to evaluate multiple interactions between genes, may be needed before genetic profiles involving this class of genes can be used for risk prediction.[14] As the computational methods for finding coding sequences embedded in a sequence of genomic DNA become increasingly powerful, the value of the human genome sequence as a tool for identifying unknown genes also increases. These algorithms for finding genes have largely replaced laborious experimental techniques to identify potential coding sequences of unknown genes for mutation analysis within linkage regions. These methods are another illustration of the fact that it is the annotation of the genomic sequence (i.e., the identification of genes and their function) that brings the sequence to life. Annotation parses the sequence into genes and noncoding regions. By including genomic features such as CpG islands, which mark the promoter regions of many genes, annotations produce a complete rendering of each sequence. Annotated sequences are publicly available in several data bases

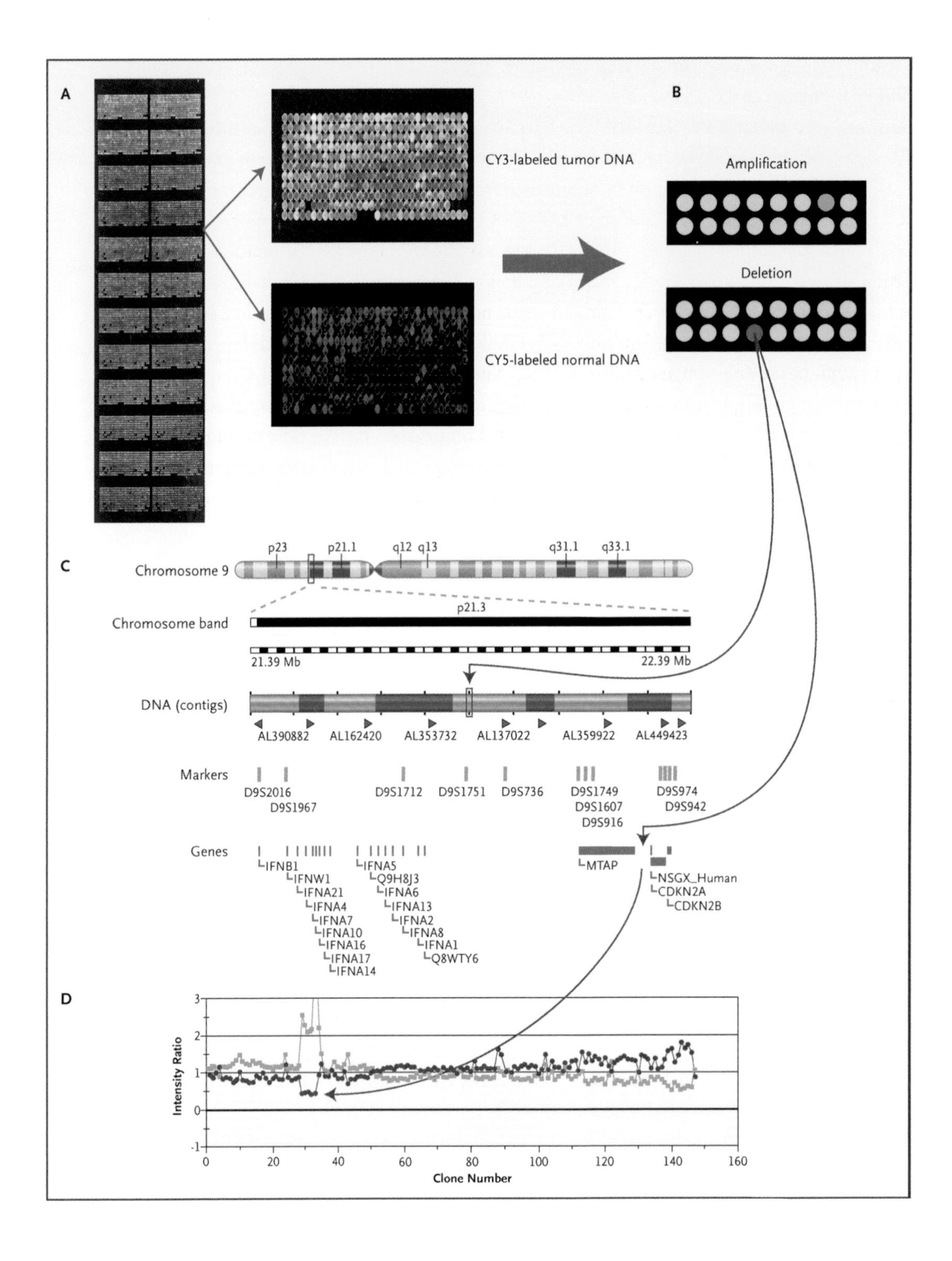
A
B
CY3-labeled tumor DNA
CY5-labeled normal DNA
Amplification
Deletion
C
Chromosome 9
p23
p21.1
q12
q13
q31.1
q33.1
Chromosome band
p21.3
21.39 Mb
22.39 Mb
DNA (contigs)
AL390882
AL162420
AL353732
AL137022
AL359922
AL449423
Markers
D9S2016
D9S1967
D9S1712
D9S1751
D9S736
D9S1749
D9S1607
D9S916
D9S974
D9S942
Genes
IFNB1
IFNW1
IFNA21
IFNA4
IFNA7
IFNA10
IFNA16
IFNA17
IFNA14
IFNA5
Q9H8J3
IFNA6
IFNA13
IFNA2
IFNA8
IFNA1
Q8WTY6
MTAP
NSGX_Human
CDKN2A
CDKN2B
D
Intensity Ratio
Clone Number

(http://www.ensembl.org, http://www.ncbi.nlm.nih.gov, and http://www.genome.ucsc.edu) with associated genome browsers.

The depth and value of annotation have also grown through the addition of millions of single-nucleotide polymorphisms, which are invaluable in the search for susceptibility genes.[26] One such example is the recent demonstration that a silent single-nucleotide polymorphism in *LIG4*, a gene encoding a DNA ligase important in the repair of breaks in double-stranded DNA, is associated with survival among patients with breast cancer.[27] This effect was demonstrated in a British population-based study that included 2430 cases of breast cancer. DNA from these patients was genotyped for polymorphisms in 22 DNA-repair, hormone-metabolism, carcinogen-metabolism, and other genes, and the effect of each single-nucleotide polymorphism on the outcome was assessed by Cox regression analysis. The silent polymorphism D501D (t > c) in *LIG4* had the largest effect. The estimated hazard ratio for death among patients homozygous for the polymorphism, as compared with those homozygous for the wild-type sequence, was 4.0 (95 percent confidence interval, 2.1 to 7.7; $P = 0.002$), and this effect remained significant after stratification according to tumor stage, grade, and type (hazard ratio, 4.2; 95 percent confidence interval, 1.8 to 9.4; $P = 0.01$). The inclusion of these single-nucleotide polymorphisms in the annotation of the human genome sequence greatly facilitated this analysis, which would otherwise have had to have been preceded by an extensive sequence-based effort to identify single-nucleotide polymorphisms.

Whole-Genome Approaches to the Analysis of Breast and Ovarian Cancer

Many genomic approaches to the identification of cancer genes are based on microarray techniques. For gene-expression profiling, each element usually represents one gene and is created with the use of a complementary DNA (cDNA) or oligonucleotide for the gene in question. Similar arrays have been produced with large genomic clones for array-based comparative genomic hybridization to identify changes in the number of copies of DNA. This approach replaced lower-resolution comparative genomic hybridization of cells in metaphase and provides a direct link to genes in the altered region (Fig. 11.3).

A comparative genomic hybridization can be used to identify the loss of one or both copies of a given gene as well as regions of amplification. Arrays made with cDNAs can be used for expression profiling and comparative genomic hybridization simultaneously.[28] This approach allows a direct comparison between the number of copies of a gene and the level of expression of that gene, but the results of comparative genomic hybridization

Figure 11.3 (facing page). Array-Based Comparative Genomic Hybridization. In Panel A, bacterial artificial chromosome clones or complementary DNAs are placed on glass slides at high density; tumor and normal DNA are labeled with CY3 and CY5, respectively; and the combined probe is hybridized to the array. The array is analyzed with use of a laser scanner that reads each color channel individually and then calculates an intensity ratio for each spot. In Panel B, spots with intensity ratios greater than 1.25 (green spots) represent increases in copy number (amplification), and those with intensity ratios of less than 0.75 (red spots) represent decreases in copy number (deletion). Each spot is a DNA segment that can be linked directly to the human genome sequence (Panel C), thus defining changes in the number of copies of a specific gene. In Panel D, the plotting of intensity ratios for the chromosome 9 bacterial artificial chromosome clones on the array in linear order identifies a homozygous loss of *CDKN2A* in a melanoma cell line.

may be variable, presumably because the cDNA sequence and the genomic sequence are not collinear. However, the alternative approach of using large cloned segments of genomic DNA in the bacterial artificial chromosomes consistently provides excellent data.[29] The genomic clones can be spaced evenly across the genome, and the array set can be enriched with selected clones that contain candidate cancer genes to enhance resolution. The use of DNA microarrays has been suggested for other applications; however, epigenetic changes such as changes in DNA methylation, which are likely to be a critical component in the development of cancer, have been notoriously difficult to assay regardless of the format.

The power of a comparative genomic hybridization was demonstrated by Albertson and colleagues, who used this approach to map the recurrent breast-cancer amplicon at chromosome 20q12.3.[30] This approach clearly demonstrated that what had previously been described as a single amplicon was, in fact, two distinct amplicons, one containing the putative oncogene *ZNF217*[31] and the other containing *CYP24*, which encodes vitamin D24-hydroxylase.[32] The overexpression of this enzyme alters the control of growth mediated by vitamin D. There were two distinct peaks of high copy numbers within this 2-Mb region, with a gene at the peak of each amplicon. The ability of comparative genomic hybridization to show peaks in increases in copy numbers across regions of recurrent abnormality at high resolution is very useful for locating oncogenes in many human cancers.

Much less advanced, but critically important, are techniques involving proteomics, which examine the entire complement of proteins expressed in a specific tissue or cell. The information supplied complements that provided by a comparative genomic hybridization, expression profiling, and screening for mutations in cancer research,[33] since the genetic code does not indicate which proteins are expressed, in what quantity, and in what form. For example, post-translational modifications, such as phosphorylation or glycosylation, may determine the function or stability of a protein and are not detected by transcriptional analyses. Many differences between normal tissue and malignant tumors are due to post-translational modifications, and a complete analysis of the cancer phenotype will require a whole-proteome approach.

Diffuse large β-cell lymphoma was the first human cancer to undergo gene-expression profiling, and a microarray containing 17,800 cDNAs was used.[34] Breast and ovarian cancer have now been subjected to molecular profiling as well. In the first such study, Perou and colleagues used a cDNA microarray containing 8000 genes to assay 65 breast-biopsy specimens, primarily invasive breast cancers.[35] Perhaps not surprisingly, estrogen-receptor status was a key predictor of the outcome and treatment response, with estrogen-receptor and coregulated genes being the primary elements needed for these tumors to cluster. In addition, a profile of tumors that overexpress *ERBB2* was easily identifiable. Thus, the primary clusters recognized were tumors that expressed estrogen receptor and had a luminal-cell pattern of gene expression, tumors that did not express estrogen receptor and had a myoepithelial-cell pattern of expression, tumors that overexpressed *ERBB2*, and a fourth group of tumors that clustered with normal breast tissue. More recently, Hedenfalk and colleagues suggested that transcriptional profiling can also accurately differentiate breast cancers with underlying germ-line mutations in *BRCA1* or *BRCA2* from those without such mutations, an advance that could facilitate the identification of high-risk families on the basis of molecular phenotyping, as well as identify characteristic molecular differences that may be useful clinical targets for directed therapy.[36]

Ovarian cancers have been subjected to transcriptional profiling with similar results. In one series, 27 serous papillary ovarian cancers and 3 samples of normal ovarian tissue underwent gene-expression profiling with oligonucleotide-based arrays representing more than 6000 human genes.[37] Normal ovarian tissue was clearly distinguishable from malignant tissues, and three types of tumors were identified. The first subtype clustered with normal tissue and was well differentiated on conventional histologic analyses. The second group was characterized by the expression of genes from admixed stromal cells and infiltrating lymphocytes. Although this profile could represent a random admixture of cell types, it could also represent an immune response to the tumor, as suggested recently by Zhang and colleagues.[38] The third group of tumors had a high level of expression of cell-cycle–associated genes, most likely reflecting a high proliferative rate. More than half the tumors in this cluster were poorly differentiated on histologic analysis. Which of the differentially expressed genes in this and other series represent the root causes of malignant transformation rather than markers of progression is not known but must be determined in order to distinguish diagnostic markers from therapeutic targets.

Microarrays have also been used to show how an expression profile can change as cancer cells develop resistance to doxorubicin-based therapy.[39] In these experiments, a set of genes that were transiently overexpressed after initial exposure to doxorubicin included a subgroup of genes that became constitutively overexpressed as resistance to doxorubicin developed. These experiments, although just the beginning of what will be a fundamental change in molecular oncology owing to the deciphering of the human genome sequence, demonstrate the power of this approach.

In perhaps the most extensive and informative study to date, the expression profiles of 117 primary breast cancers were compared with known prognostic markers and the clinical outcome at least five years after diagnosis.[40] Expression profiling with the use of 25,000 genes separated the tumors into two groups, one in which distant metastases developed in 34 percent at five years and one in which metastases developed in 70 percent at five years. From the original 25,000 genes in the array, 70 were identified as having the greatest accuracy in predicting recurrent disease. When the tumors were sorted on the basis of this smaller set of genes, fewer than 10 percent of the tumors in the poor-prognosis group were misclassified. A comparative multivariate analysis using clinical prognostic factors that included tumor grade, tumor size, the presence or absence of angiolymphatic invasion, patients' age, and tumor estrogen-receptor status demonstrated that as compared with the good-prognosis gene-expression signature, the poor-prognosis microarray profile was an independent predictor of recurrence, with an odds ratio of 18 (95 percent confidence interval, 3 to 94). This approach has now been tested in 295 consecutive patients with stage I or II breast cancer.[41] Of these, 180 had the poor-prognosis profile and 115 had the good-prognosis profile. Ten years after the diagnosis of breast cancer, the probability of remaining free of metastases was 51 percent among women with a poor-prognosis profile and 85 percent among those with a good-prognosis profile. These data provide compelling evidence that the genetic program of a cancer cell at diagnosis defines its biologic behavior many years later, refuting a competing hypothesis that the genetic changes driving the development of metastatic disease are acquired in residual cells after adjuvant treatment.

CLINICAL MANAGEMENT OF INHERITED SUSCEPTIBILITY TO BREAST AND OVARIAN CANCER

Several computational models have been developed to predict an individual woman's risk of breast cancer, including one in which family history is the predominant risk factor. This model, developed by Claus and colleagues and published as a series of tables clinicians can use,[42] is based on the number and degree of relatedness of family members with breast cancer and their age at diagnosis. However, this model does not provide estimates of the likelihood that an individual woman will have a germ-line mutation in *BRCA1* or *BRCA2*. Several studies have identified factors that are associated with an increased likelihood that a *BRCA1* or *BRCA2* mutation will be identified, including early-onset breast cancer, the occurrence of breast and ovarian cancer in the same woman, a history of male family members with breast cancer, and Ashkenazi Jewish ancestry. These characteristics have also been included in predictive models designed for use by clinicians.[43–45]

Testing for germ-line mutations in *BRCA1* and *BRCA2* is an important tool for predicting the risk of breast cancer and developing management strategies. Once such mutations are identified, we recommend that the woman choose between annual screening mammography and prophylactic mastectomy, which significantly reduced the risk of breast cancer in a small, retrospective study of mutation carriers.[46] We recommend that women who choose surveillance also investigate the possibility of participating in a clinical trial evaluating the utility of magnetic resonance imaging for screening high-risk women. Several studies have shown that in women with germ-line *BRCA1* and *BRCA2* mutations, breast cancers are likely to occur as interval cancers[47] and that standard mammograms are more likely to be negative than in women at low or moderate risk.[48–50]

With respect to the risk of ovarian cancer among carriers of *BRCA1* and *BRCA2* mutations, we strongly recommend that such women undergo prophylactic oophorectomy as soon as they have completed childbearing, since no surveillance regimen to date has been shown to decrease the percentage of women who receive a diagnosis of advanced disease. Among mutation carriers, this procedure has been shown to reduce the risk of breast and ovarian cancer by more than 60 percent and 95 percent, respectively.[51,52] Again, although no prospective data are available, we recommend that these women receive hormone-replacement therapy until the age of 50 years, approximately the time of natural menopause. Although extending exposure to estrogens beyond the age of 50 years has been associated with a small increase in the risk of breast cancer, these younger women would be producing endogenous estrogens in the absence of prophylactic oophorectomy.

The addition of hormone-replacement therapy makes this choice acceptable to women who would otherwise refuse it because of concern about premature menopause, and the risk–benefit ratio is strongly in favor of oophorectomy, with or without hormone-replacement therapy.

The use of tamoxifen to prevent breast cancer in carriers of *BRCA1* and *BRCA2* mutations remains controversial. One retrospective study suggested that adjuvant tamoxifen therapy in carriers of *BRCA1* and *BRCA2* mutations with estrogen-receptor–positive breast cancer reduced the risk of contralateral breast cancers by the same amount as that in unselected patients with breast cancer.[53] However, data showing that most *BRCA1*-associated breast cancers are negative for estrogen receptors[54] and recent data from the Breast Cancer Prevention Trial (BCPT) of the National Surgical Adjuvant Breast and Bowel Project have led to widespread speculation that tamoxifen will not prevent breast cancer in women with germ-line *BRCA1* mutations.[55] It is important to consider that all available data suggest that endogenous exposure to hormones has a central role in defining the risk of cancer among carriers of *BRCA1* mutations, that breast cancer developed in only eight carriers of *BRCA1* mutations in the BCPT, and that the lack of a preventive effect of tamoxifen was statistically insignificant (odds ratio, 1.67; 95 percent confidence interval, 0.41 to 8.00). The length of treatment in the BCPT is also consistent with an early treatment effect, rather than true prevention, which, given the preponderance of estrogen-receptor–negative tumors among carriers of *BRCA1* mutations, would produce the data seen in this study. Thus, we recommend that carriers of *BRCA1* mutations consider taking tamoxifen once they discontinue hormone-replacement therapy at about the age of 50 years.

SUMMARY

The past decade has been a period of unparalleled discovery in the field of the genetics and genomics of breast and ovarian cancer. Two major susceptibility genes have been isolated, and subsequent work provided sufficient management information to allow genetic testing for *BRCA1* and *BRCA2* mutations to become a part of routine practice in many clinical centers. In addition, work has begun on the characterization of genetic variants that, although associated with a lower risk of cancer than germ-line *BRCA1* and *BRCA2* mutations, are far more common in the population and thus may have a substantial role in defining the risk of cancer. Finally, gene-expression profiling, coupled with the sequencing of most or all of the genes in the human genome, is revolutionizing the study of the biology and the molecular classification of breast and ovarian cancer. Combined with data from projects conducting a genome-wide mutation analysis of all genes implicated in the development of cancer, the importance of which has just been illustrated with the discovery that more than 60 percent of melanomas have mutations in *BRAF* (v-raf murine sarcoma viral oncogene homologue B1),[56] and progress in developing effective preventive measures, a marked reduction in mortality from breast and ovarian cancer is a realistic goal for the next decade.

References

1. Madigan MP, Ziegler RG, Benichou J, Byrne C, Hoover RN. Proportion of breast cancer cases in the United States explained by well-established risk factors. J Natl Cancer Inst 1995; 87:1681–5.

2. Miki Y, Swensen J, Shattuck-Eidens D, et al. A strong candidate for the breast and ovarian cancer susceptibility gene BRCA1. Science 1994; 266:66–71.

3. Wooster R, Bignell G, Lancaster J, et al. Identification of the breast cancer susceptibility gene BRCA2. Nature 1995; 378:789–92. [Erratum, Nature 1996; 379:749.]

4. Brose MS, Rebbeck TR, Calzone KA, Stopfer JE, Nathanson KL, Weber BL. Cancer risk estimates for BRCA1 mutation carriers identified in a risk evaluation program. J Natl Cancer Inst 2002; 94: 1365–72.

5. Thompson D, Easton DF. Cancer incidence in BRCA1 mutation carriers. J Natl Cancer Inst 2002; 94:1358–65.

6. Newman B, Millikan RC, King MC. Genetic epidemiology of breast and ovarian cancers. Epidemiol Rev 1997; 19:69–79.

7. Ford D, Easton DF, Peto J. Estimates of the gene frequency of BRCA1 and its contribution to breast and ovarian cancer incidence. Am J Hum Genet 1995; 57:1457–62.

8. Garber JE, Goldstein AM, Kantor AF, Dreyfus MG, Fraumeni JF Jr, Li FP. Follow-up study of twenty-four families with Li-Fraumeni syndrome. Cancer Res 1991; 51:6094–7.

9. Meijers-Heijboer H, van den Ouweland A, Klijn J, et al. Low-penetrance susceptibility to breast cancer due to CHEK2(*) 1100delC in noncarriers of BRCA1 or BRCA2 mutations. Nat Genet 2002; 31:55–9.

10. Chenevix-Trench G, Spurdle AB, Gatei M, et al. Dominant negative ATM mutations in breast cancer families. J Natl Cancer Inst 2002; 94:205–15. [Erratum, J Natl Cancer Inst 2002; 94:952.]

11. Spring K, Ahangari F, Scott SP, et al. Mice heterozygous for mutation in Atm, the gene involved in ataxia-telangiectasia, have heightened susceptibility to cancer. Nat Genet 2002; 32:185–90.

12. Gayther SA, Russell P, Harrington P, Antoniou AC, Easton DF, Ponder BAJ. The contribution of germline BRCA1 and BRCA2 mutations to familial ovarian cancer: no evidence for other ovarian cancer-susceptibility genes. Am J Hum Genet 1999; 65:1021–9.

13. Sekine M, Nagata H, Tsuji S, et al. Localization of a novel susceptibility gene for familial ovarian cancer to chromosome 3p22-p25. Hum Mol Genet 2001; 10:1421–9.

14. Dunning AM, Healey CS, Pharoah PD, Teare MD, Ponder BA, Easton DF. A systematic review of genetic polymorphisms and breast cancer risk. Cancer Epidemiol Biomarkers Prev 1999; 8:843–54.

15. Hall JM, Lee MK, Newman B, et al. Linkage of early-onset familial breast cancer to chromosome 17q21. Science 1990; 250:1684–9.

16. Wooster R, Neuhausen SL, Mangion J, et al. Localization of a breast cancer susceptibility gene, BRCA2, to chromosome 13q12-13. Science 1994; 265:2088–90.

17. Schutte M, da Costa LT, Hahn SA, et al. Identification by representational difference analysis of a homozygous deletion in pancreatic carcinoma that lies within the BRCA2 region. Proc Natl Acad Sci U S A 1995; 92:5950–4.

18. Barkardottir RB, Arason A, Egilsson V, Gudmundsson J, Jonasdottir A, Johannesdottir G. Chromosome 17q-linkage seems to be infrequent in Icelandic families at risk of breast cancer. Acta Oncol 1995; 34:657–62.

19. Gudmundsson J, Johannesdottir G, Arason A, et al. Frequent occurrence of BRCA2 linkage in Icelandic breast cancer families and segregation of a common BRCA2 haplotype. Am J Hum Genet 1996; 58:749–56.

20. Nyström-Lahti M, Peltomäki P, Aaltonen LA, de la Chapelle A. Genetic and genealogic study of 33 Finnish HNPCC kindreds. Am J Hum Genet 1994; 55:Suppl:A365. abstract.

21. Kainu T, Juo SH, Desper R, et al. Somatic deletions in hereditary breast cancers implicate 13q21 as a putative novel breast cancer susceptibility locus. Proc Natl Acad Sci U S A 2000; 97:9603–8.

22. Seitz S, Rohde K, Bender E, et al. Deletion mapping and linkage analysis provide strong indication for the involvement of the human chromosome region 8p12-p22 in breast carcinogenesis. Br J Cancer 1997; 76:983–91.

23. Rahman N, Teare MD, Seal S, et al. Absence of evidence for a familial breast cancer susceptibility gene at chromosome 8p12-p22. Oncogene 2000; 19:4170–3.

24. Thompson D, Szabo CI, Mangion J, et al. Evaluation of linkage of breast cancer to the putative BRCA3 locus on chromosome 13q21 in 128 multiple case families from the Breast Cancer Linkage Consortium. Proc Natl Acad Sci U S A 2002; 99:827–31.

25. Ford D, Easton DF, Stratton M, et al. Genetic heterogeneity and penetrance analysis of the BRCA1 and BRCA2 genes in breast cancer families. Am J Hum Genet 1998; 62:676–89.

26. Sachidanandam R, Weissman D, Schmidt SC, et al. A map of human genome sequence variation containing 1.42 million single nucleotide polymorphisms. Nature 2001; 409:928–33.

27. Goode EL, Dunning AM, Kuschel B, et al. Effect of germ-line genetic variation on breast cancer survival in a population-based study. Cancer Res 2002; 62:3052–7.

28. Pollack JR, Sorlie T, Perou CM, et al. Microarray analysis reveals a major direct role of DNA copy number alteration in the transcriptional program of human breast tumors. Proc Natl Acad Sci U S A 2002; 99:12963–8.

29. Snijders AM, Nowak N, Segraves R, et al. Assembly of microarrays for genome-wide measurement of DNA copy number. Nat Genet 2001; 29:263–4.

30. Albertson DG, Ylstra B, Segraves R, et al. Quantitative mapping of amplicon structure by array CGH identifies CYP24 as a candidate oncogene. Nat Genet 2000; 25:144–6.

31. Collins C, Rommens JM, Kowbel D, et al. Positional cloning of ZNF217 and NABC1: genes amplified at 20q13.2 and overexpressed in breast carcinoma. Proc Natl Acad Sci U S A 1998; 95:8703–8.

32. Walters MR. Newly identified actions of the vitamin D endocrine system. Endocr Rev 1992; 13:719–64.

33. Figeys D, Gygi SP, Zhang Y, Watts J, Gu M, Aebersold R. Electrophoresis combined with novel mass spectrometry techniques: powerful tools for the analysis of proteins and proteomes. Electrophoresis 1998; 19:1811–8.

34. Alizadeh AA, Eisen MB, Davis RE, et al. Distinct types of diffuse large B-cell lymphoma identified by gene expression profiling. Nature 2000; 403:503–11.

35. Perou CM, Jeffrey SS, van de Rijn M, et al. Distinctive gene expression patterns in human mammary epithelial cells and breast cancers. Proc Natl Acad Sci U S A 1999; 96:9212–7.

36. Hedenfalk I, Duggan D, Chen Y, et al. Gene-expression profiles in hereditary breast cancer. N Engl J Med 2001; 344:539–48.

37. Welsh JB, Zarrinkar PP, Sapinoso LM, et al. Analysis of gene expression profiles in normal and neoplastic ovarian tissue samples identifies candidate molecular markers of epithelial ovarian cancer. Proc Natl Acad Sci U S A 2001; 98:1176–81.

38. Zhang L, Conejo-Garcia JR, Katsaros D, et al. Intratumoral T cells, recurrence, and survival in epithelial ovarian cancer. N Engl J Med 2003; 348:203–13.

39. Kudoh K, Ramanna M, Ravatn R, et al. Monitoring the expression profiles of doxorubicin-induced and doxorubicin-resistant cancer cells by cDNA microarray. Cancer Res 2000; 60:4161–6.

40. van 't Veer LJ, Dai H, van de Vijver MJ, et al. Gene expression profiling predicts clinical outcome of breast cancer. Nature 2002; 415:530–6.

41. van de Vijver MJ, He YD, van 't Veer LJ, et al. A gene-expression signature as a predictor of survival in breast cancer. N Engl J Med 2002; 347:1999–2009.

42. Claus EB, Risch N, Thompson WD. Genetic analysis of breast cancer in the cancer and steroid hormone study. Am J Hum Genet 1991; 48:232–42.

43. Couch FJ, DeShano ML, Blackwood MA, et al. *BRCA1* mutations in women attending clinics that evaluate the risk of breast cancer. N Engl J Med 1997; 336:1409–15.

44. Frank TS, Manley SA, Olopade OI, et al. Sequence analysis of BRCA1 and BRCA2: correlation of mutations with family history and ovarian cancer risk. J Clin Oncol 1998; 16:2417–25.

45. Parmigiani G, Berry D, Aguilar O. Determining carrier probabilities for breast cancer-susceptibility genes BRCA1 and BRCA2. Am J Hum Genet 1998; 62:145–58.

46. Meijers-Heijboer H, van Geel B, van Putten WLJ, et al. Breast cancer after prophylactic bilateral mastectomy in women with a *BRCA1* or *BRCA2* mutation. N Engl J Med 2001; 345:159–64.

47. Brekelmans CT, Seynaeve C, Bartels CC, et al. Effectiveness of breast cancer surveillance in BRCA1/2 gene mutation carriers and women with high familial risk. J Clin Oncol 2001; 19:924–30.

48. Warner E, Plewes DB, Shumak RS, et al. Comparison of breast magnetic resonance imaging, mammography, and ultrasound for surveillance of women at high risk for hereditary breast cancer. J Clin Oncol 2001; 19:3524–31.

49. Tilanus-Linthorst M, Verhoog L, Obdeijn IM, et al. A BRCA1/2 mutation, high breast density and prominent pushing margins of a tumor independently contribute to a frequent false-negative mammography. Int J Cancer 2002; 102:91–5. [Erratum, Int J Cancer 2002; 102:665.]

50. Stoutjesdijk MJ, Boetes C, Jager GJ, et al. Magnetic resonance imaging and mammography in women with a hereditary risk of breast cancer. J Natl Cancer Inst 2001; 93:1095–102.

51. Rebbeck TR, Levin AM, Eisen A, et al. Breast cancer risk after bilateral prophylactic oophorectomy in BRCA1 mutation carriers. J Natl Cancer Inst 1999; 91:1475–9.

52. Rebbeck TR, Lynch HT, Neuhausen SL, et al. Prophylactic oophorectomy in carriers of *BRCA1* or *BRCA2* mutations. N Engl J Med 2002; 346:1616–22.

53. Narod SA, Brunet JS, Ghadirian P, et al. Tamoxifen and risk of contralateral breast cancer in BRCA1 and BRCA2 mutation carriers: a case-control study. Lancet 2000; 356:1876–81.

54. Armes JE, Trute L, White D, et al. Distinct molecular pathogeneses of early-onset breast cancers in BRCA1 and BRCA2 mutation carriers: a population-based study. Cancer Res 1999; 59:2011–7.

55. King MC, Wieand S, Hale K, et al. Tamoxifen and breast cancer incidence among women with inherited mutations in BRCA1 and BRCA2: National Surgical Adjuvant Breast and Bowel Project (NSABP-P1) Breast Cancer Prevention Trial. JAMA 2001; 286:2251–6.

56. Davies H, Bignell GR, Cox C, et al. Mutations of the BRAF gene in human cancer. Nature 2002; 417:949–54.

12

Cardiovascular Disease

ELIZABETH G. NABEL, M.D.

Cardiovascular disease, including stroke, is the leading cause of illness and death in the United States. There are an estimated 62 million people with cardiovascular disease and 50 million people with hypertension in this country.[1] In 2000, approximately 946,000 deaths were attributable to cardiovascular disease, accounting for 39 percent of all deaths in the United States.[2] Epidemiologic studies and randomized clinical trials have provided compelling evidence that coronary heart disease is largely preventable.[3] However, there is also reason to believe that there is a heritable component to the disease. In this chapter, I highlight what we know now about genetic factors in cardiovascular disease. As future genomic discoveries are translated to the care of patients with cardiovascular disease, it is likely that what we can do will change.

LESSONS LEARNED FROM MONOGENIC CARDIOVASCULAR DISORDERS

Our understanding of the mechanism by which single genes can cause disease, even though such mechanisms are uncommon, has led to an understanding of the pathophysiological basis of more common cardiovascular diseases, which clearly are genetically complex. This point can be illustrated by a description of the genetic basis of specific diseases.

Elevated Levels of Low-Density Lipoprotein Cholesterol and Coronary Artery Disease

Low-density lipoprotein (LDL) is the major cholesterol-carrying lipoprotein in plasma and is the causal agent in many forms of coronary heart disease (Fig. 12.1). Four monogenic diseases elevate plasma levels of LDL by impairing the activity of hepatic LDL receptors, which normally clear LDL from the plasma (Table 12.1). Familial hypercholesterolemia was the first monogenic disorder shown to cause elevated plasma cholesterol levels. The primary

Originally published July 3, 2003

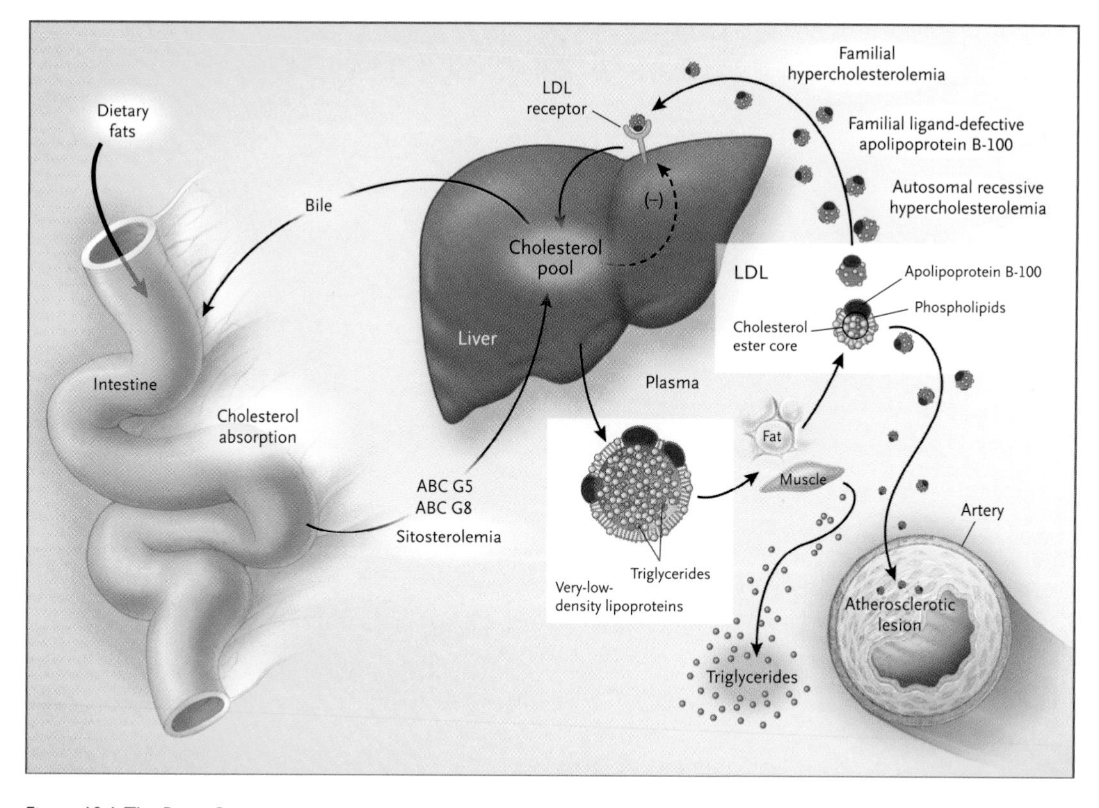

Figure 12.1 The Basic Components of Cholesterol Synthesis and Excretion. Low-density lipoprotein (LDL) molecules are composed of a cholesteryl ester core surrounded by a coat made up of phospholipid and apolipoprotein B-100. The liver secretes LDLs as larger precursor particles called very-low-density lipoproteins, which contain triglycerides and cholesterol esters. Capillaries in muscle and adipose tissue remove the triglycerides, and the lipid particle is modified into an LDL, with its cholesteryl ester core and apolipoprotein B-100 coat. LDLs circulate in the plasma, and the apolipoprotein B-100 component binds to LDL receptors on the surface of hepatocytes. Through receptor-mediated endocytosis, receptor-bound LDLs enter hepatocytes and undergo degradation in lysosomes, and the cholesterol remnants enter a cellular cholesterol pool. A negative-feedback loop regulates the number of LDL receptors. A rise in the hepatocyte cholesterol level suppresses the transcription of LDL-receptor genes, and LDL is retained in the plasma. Conversely, a decrease in hepatic cholesterol stimulates the transcription of LDL-receptor genes, removing LDL from the plasma. This mechanism accounts for the LDL-lowering action of the statins, which inhibit an enzymatic step in hepatic cholesterol synthesis. Four monogenetic diseases that elevate plasma LDL are highlighted in yellow. ABC denotes ATP-binding cassette.

defect in familial hypercholesterolemia is a deficit of LDL receptors, and more than 600 mutations in the *LDLR* gene have been identified in patients with this disorder.[5] One in 500 people is heterozygous for at least one such mutation, whereas only 1 in a million is homozygous at a single locus. Those who are heterozygous produce half the normal number of LDL receptors, leading to an increase in plasma LDL levels by a factor of 2 or 3, whereas LDL levels in those who are homozygous are 6 to 10 times normal levels. Homozygous persons have severe coronary atherosclerosis and usually die in childhood from myocardial infarction.

Deficiency of lipoprotein transport abolishes transporter activity, resulting in elevated cholesterol absorption and LDL synthesis. For example, mutations in the *APOB-100* gene, which encodes apolipoprotein B-100, reduce the binding of apolipoprotein B-100 to LDL receptors and slow the clearance of plasma LDL, causing a disorder known as

Table 12.1 Monogenic Diseases That Elevate Plasma Levels of Low-Density Lipoprotein (LDL) Cholesterol*

Disease	Mutant Gene	Molecular Mechanism	Approximate Plasma Cholesterol Level† (mg/dl)
Familial hypercholesterolemia	*LDLR*	Nonfunctional receptor fails to take up plasma cholesterol	
Homozygous			650
Heterozygous			300
Familial ligand-defective apolipoprotein B-100	*APOB-100*	Apolipoprotein B-100 fails to bind LDL receptor	
Homozygous			325
Heterozygous			275
Autosomal recessive hypercholesterolemia	*ARH*	LDL-receptor activity is disrupted	450
Sitosterolemia	*ABCG5* and *ABCG8*	Transcription factors (liver X receptor and sterol regulatory element binding protein) that regulate liver cholesterol synthesis and clearance are suppressed	150–650

*The information is adapted from Goldstein and Brown,[4] with the permission of the publisher.

†To convert the values for LDL cholesterol to millimoles per liter, multiply by 0.02586.

familial ligand-defective apolipoprotein B-100.[6] One in 1000 people is heterozygous for one of these mutations; lipid profiles and clinical disease in such persons are similar to those of persons heterozygous for a mutation causing familial hypercholesterolemia.

Sitosterolemia, a rare autosomal disorder, results from loss-of-function mutations in genes encoding two ATP-binding cassette (ABC) transporters, ABC G5 and ABC G8,[7,8] which act in concert to export cholesterol into the intestinal lumen, thereby diminishing cholesterol absorption. Autosomal recessive hypercholesterolemia is extremely rare (prevalence, <1 case per 10 million persons). The molecular cause is the presence of defects in a putative hepatic adaptor protein, which then fails to clear plasma LDL with LDL receptors.[9] Mutations in the gene encoding that protein (*ARH*) elevate plasma LDL to levels similar to those seen in homozygous familial hypercholesterolemia.

In the majority of people with hypercholesterolemia in the general public, the condition is attributable to high-fat diets and poorly understood susceptibility and modifier genes. Study of the monogenic disorders, noted above, that disrupt LDL-receptor pathways has clarified the importance of cholesterol synthesis and excretion pathways in the liver and has highlighted molecular targets for regulating plasma cholesterol levels. For example, statin therapy for hypercholesterolemia is based on an understanding of the molecular basis of that disorder.

Hypertension

Hypertension is the most common disease in industrialized nations, with a prevalence above 20 percent in the general population. It imparts an increased risk of stroke, myocardial infarction, heart failure, and renal failure; many clinical trials have shown that reductions in blood pressure reduce the incidence of stroke and myocardial infarction.[10] Multiple environmental and genetic determinants complicate the study of blood-pressure variations in the general population. In contrast, the investigation of rare mendelian forms of blood-pressure variation in which mutations in single genes cause marked extremes in blood pressure has been very informative (Table 12.2). These mutations, which impair renal salt handling, provide a molecular basis for understanding the pathogenesis of hypertension (Fig. 12.2).[11]

Investigation of families with severe hypertension or hypotension has identified mutations in genes that regulate these pathways. Pseudohypoaldosteronism type II is an autosomal dominant disorder characterized by hypertension, hyperkalemia, increased renal salt reabsorption, and impaired potassium- and hydrogen-ion excretion. Wilson and colleagues identified two genes causing pseudohypoaldosteronism type II; both encode proteins in the WNK family of serine–threonine kinases.[12] Mutations in *WNK1* are intronic deletions on chromosome 12p. Missense mutations in *hWNK4*, on chromosome 17, also cause pseudohypoaldosteronism type II. Immunofluorescence assays have shown that the proteins localize to distal nephrons and may serve to increase transcellular chloride conductance in the collecting ducts, leading to salt reabsorption, increased intravascular volume, and diminished secretion of potassium and hydrogen ions.

Abnormalities in the activity of aldosterone synthase produce hypertension or hypotension. Glucocorticoid-remediable aldosteronism is an autosomal dominant trait featuring early-onset hypertension with suppressed renin activity and normal or elevated aldosterone levels. This form of aldosteronism is caused by gene duplication arising from an unequal crossover between two genes that encode enzymes in the adrenal-steroid biosynthesis pathway (aldosterone synthase and 11β-hydroxylase).[13,14] The chimeric gene encodes a protein with aldosterone synthase activity that is ectopically expressed in the adrenal fasciculata under the control of corticotropin rather than angiotensin II. Normal cortisol production leads to constitutive aldosterone secretion, plasma-volume expansion, hypertension, and suppressed renin levels. Mutations that cause a loss of aldosterone synthase activity impair renal salt retention and the secretion of potassium and hydrogen ions in the distal nephrons and lead to severe hypotension as a result of reduced intravascular volume.[15]

Mutations that alter renal ion channels and transporters give rise to Liddle's, Gitelman's, and Bartter's syndromes. Liddle's syndrome is an autosomal dominant trait characterized by early-onset hypertension, hypokalemic alkalosis, suppressed renin activity, and low plasma aldosterone levels due to mutations in the epithelial sodium channel.[16,17] Loss-of-function mutations in the gene encoding the thiazide-sensitive sodium–chloride cotransporter in the distal convoluted tubules cause Gitelman's syndrome.[18] Patients present in adolescence or early adulthood with neuromuscular signs and symptoms, a lower than normal blood pressure, a low serum magnesium level, and a low urinary calcium level. Bartter's syndrome can be produced by mutations in any of three genes required for normal salt reabsorption in the thick ascending loop of Henle; it can be distinguished from Gitelman's syndrome because it features increased urinary calcium levels and normal or reduced magnesium levels.[19] In these inherited disorders, the net salt balance consistently predicts the blood pressure. As a result, new targets for antihypertensive therapy, including the epithelial sodium channel, other ion channels, and the WNK kinases, have been identified.

Table 12.2 Monogenic Diseases That Elevate or Lower Blood Pressure*

Disease	Mutation	Molecular Mechanism	Effect on Blood Pressure
Glucocorticoid-remediable aldosteronism	Duplication of genes encoding aldosterone synthase and 11β-hydroxylase, caused by an unequal crossover	Ectopic expression of a protein with aldosterone synthase activity regulated by corticotropin; increased plasma volume	Increased
Aldosterone synthase deficiency	Mutations in the gene encoding aldosterone synthase	Defective aldosterone synthase activity; decreased plasma volume	Decreased
21-Hydroxylase deficiency	Mutations in the gene encoding 21-hydroxylase	Absence of circulating aldosterone; decreased plasma volume	Decreased
Apparent mineralocorticoid excess	Mutation in the gene encoding 11 β-hydroxysteroid dehydrogenase	Cortisol-mediated activation of the mineralocorticoid receptor, sodium retention, and expanded plasma volume	Increased
Hypertension exacerbated by pregnancy	Mutation in the ligand-binding domain of the mineralocorticoid receptor	Activation of the mineralocorticoid receptor by steroids lacking 21-hydroxyl groups (probably due in part to the rise in progesterone levels during pregnancy)	Increased
Pseudohypoaldosteronism type I (autosomal dominant)	Loss-of-function mutations in mineralocorticoid receptor	Partial loss of function of the mineralocorticoid receptor, impairing salt reabsorption; improvement with age and a high-salt diet	Decreased
Liddle's syndrome	Mutations in the ENaC β or γ subunit	Deletion of the C-terminal domain of ENaC, resulting in increased ENaC activity	Increased
Pseudohypoaldosteronism type I (autosomal recessive)	Loss-of-function mutations in ENaC subunits	Impairment of ENaC subunits, which is not ameliorated by activation of the mineralocorticoid receptor by aldosterone; no improvement with age; massive salt supplementation required	Decreased
Gitelman's syndrome	Loss-of-function mutations in the sodium–chloride cotransporter of the distal convoluted tubule	Salt wasting from the distal convoluted tubule, leading to activation of the renin-angiotensin system; subsequent activation of the mineralocorticoid receptor increases ENaC activity, preserving salt homeostasis	Normal or decreased
Bartter's syndrome	Loss-of-function mutations in genes required for salt reabsorption in the thick ascending loop of Henle	Salt wasting in the thick ascending loop of Henle leads to activation of the renin-angiotensin system and the mineralocorticoid receptor, increased ENaC activity, and relative salt homeostasis	Normal or decreased

*ENaC denotes epithelial sodium channel.

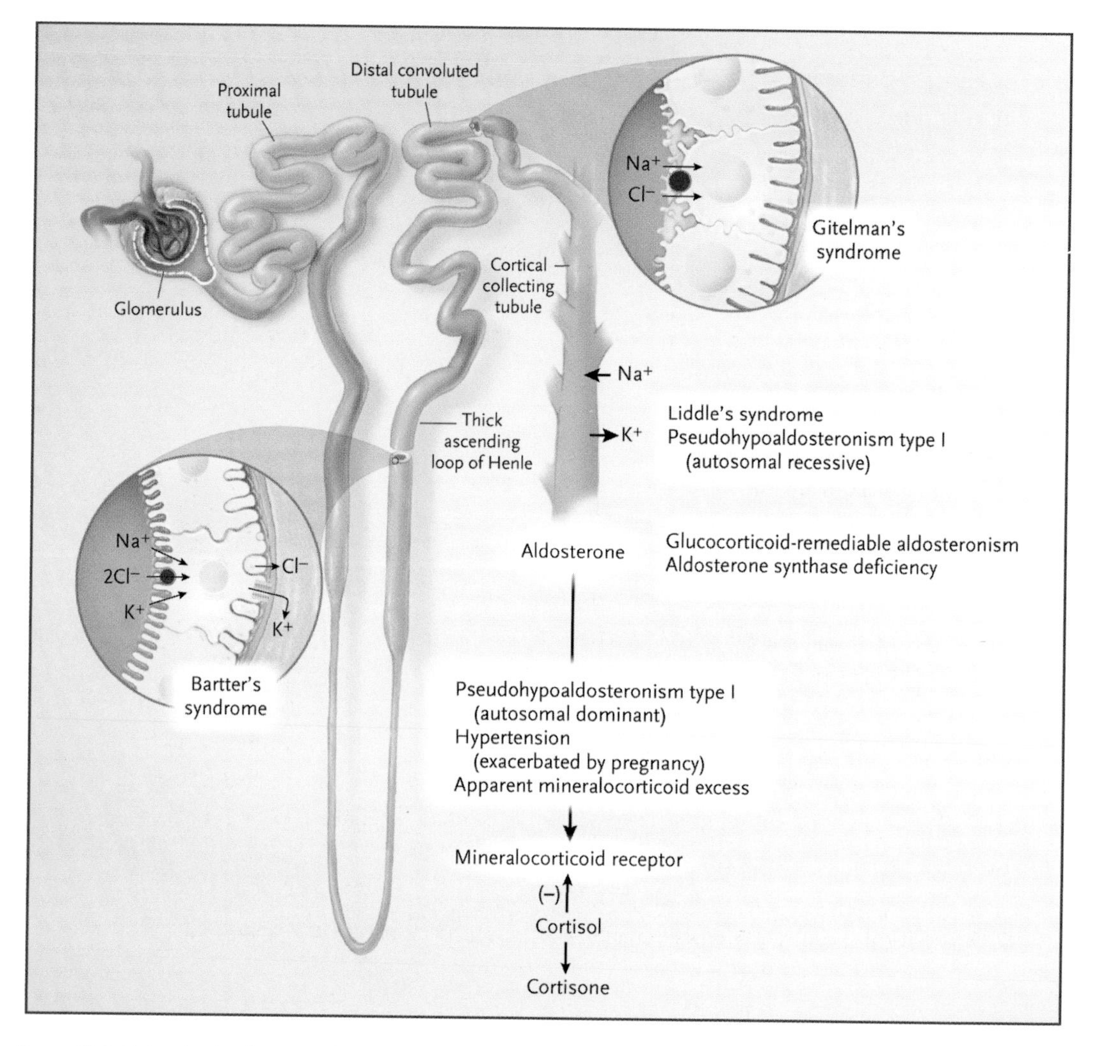

Figure 12.2 Molecular Mechanisms Mediating Salt Reabsorption in the Kidney and Associated Monogenic Hypertensive Diseases. The kidney filters more than 180 liters of plasma (containing 23 moles of salt) daily and reabsorbs more than 99 percent of the filtered sodium. The proximal tubule of the nephron reabsorbs about 60 percent of the filtered sodium, primarily by sodium–hydrogen ion exchange. The thick ascending loop of Henle absorbs about 30 percent by sodium–potassium-chloride (Na^+–K^+–$2Cl^-$) cotransporters. The distal convoluted tubule reabsorbs about 7 percent by sodium–chloride cotransporters, and the remaining 3 percent of the filtered sodium is handled by epithelial sodium channels in the cortical collecting tubule. The renin–angiotensin system tightly regulates the activity of the epithelial sodium channels. Decreased delivery of sodium to the loop of Henle leads to renin secretion by the juxtaglomerular apparatus of the kidney. Renin acts on the circulating precursor angiotensinogen to generate angiotensin I, which is converted in the lungs to angiotensin II by angiotensin-converting enzyme. Angiotensin II binds to its specific receptor in the adrenal glomerulosa, stimulating aldosterone secretion. Aldosterone binds to its receptor in the distal nephron, leading to increased activity of the epithelial sodium channels and sodium reabsorption. Monogenetic diseases that alter blood pressure are shown in yellow. (Adapted from Lifton et al.,[11] with the permission of the publisher.)

Thrombosis and Hemostasis

The blood-clotting system requires precise control of factors within and outside the coagulation cascade to prevent fatal bleeding or unwanted thrombosis. A common variant in the factor V gene, one encoding the substitution of glutamine for arginine at position 506 (Arg506Gln), prevents the degradation of factor V and promotes clot formation. This substitution, also known as factor V Leiden, has an allele frequency of 2 to 7 percent in European populations and has been observed in 20 to 50 percent of patients with venous thromboembolic disease.[20–22] Factor V Leiden has incomplete penetrance and variable expression. Approximately 80 percent of persons who are homozygous for the mutation and 10 percent of those who are heterozygous will have thrombosis at some point in their lifetime.[23,24] Factor V Leiden increases the risk of venous thrombosis in men, but not myocardial infarction or stroke.[25] In a subgroup of patients, thrombosis is associated with coinheritance of gene mutations that modify the factor V Leiden phenotype.[26–29] Identification of gene modifiers is an area of active research and is essential for distinguishing, among persons who are heterozygous for factor V Leiden, the 10 percent in whom serious thrombosis will develop from the 90 percent who will have no symptoms.

Hypertrophic Cardiomyopathy

Hypertrophic cardiomyopathy is the most common monogenic cardiac disorder and the most frequent cause of sudden death from cardiac causes in children and adolescents.[30] On the basis of the evaluation of echocardiograms from a large population of young persons, the incidence of hypertrophic cardiomyopathy has been estimated at approximately 1 in 500 persons.[31] Hypertrophic cardiomyopathy is transmitted in an autosomal dominant pattern. Mutations in the genes encoding proteins of the myocardial-contractile apparatus cause the disease (Fig. 12.3).[32] Investigators have found multiple causative mutations in at least 10 different sarcomeric proteins,[33] including cardiac β-myosin heavy chain, cardiac myosin-binding protein, cardiac troponin T, cardiac troponin I, α-tropomyosin, essential and regulatory light chains, and cardiac actin.

The pathologic features of hypertrophic cardiomyopathy consist of marked left ventricular hypertrophy, a thickened ventricular septum, atrial enlargement, and a small left

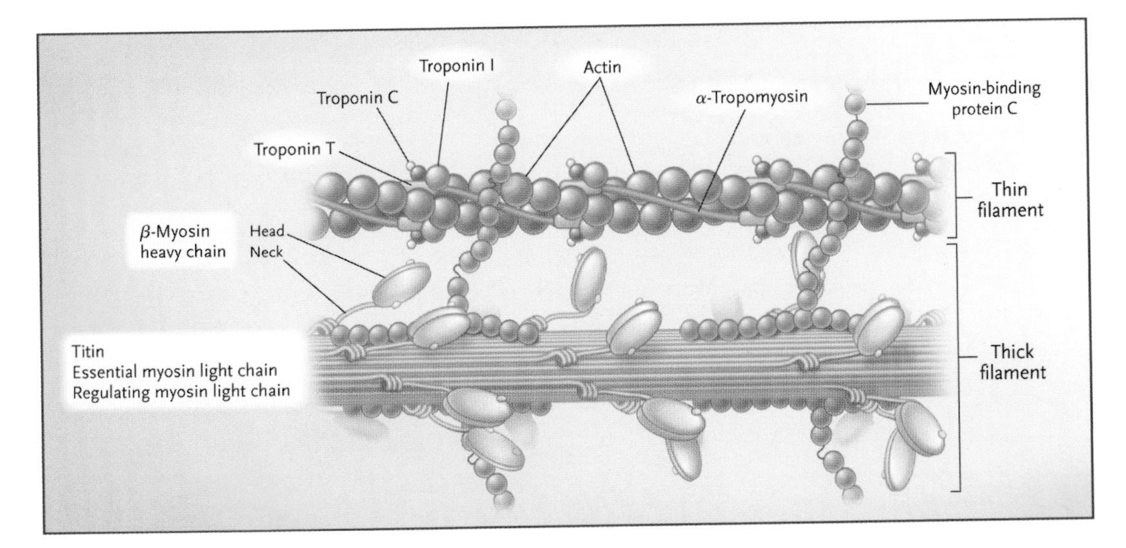

Figure 12.3 Mutations in Cardiac Sarcomeric Proteins That Cause Hypertrophic Cardiomyopathy. Sarcomeric proteins that constitute the thick and thin filaments are shown. Sarcomere proteins that cause hypertrophic cardiomyopathy are labeled in yellow. (Adapted from Kamisago et al.[32])

ventricular cavity. Hypertrophy and disarray of the myocytes and interstitial fibrosis are present throughout the myocardium. The cardiac phenotype and clinical course of patients with hypertrophic cardiomyopathy are highly variable with regard to the pattern and degree of hypertrophy, the age at onset, and the clinical outcome. This variability is due partly to the different functions performed by mutant sarcomeric proteins. For example, a mutation in the gene encoding β-myosin heavy chain was the first mutation identified as a cause of familial hypertrophic cardiomyopathy,[34] and more than 100 disease-causing mutations have since been detected.[35]

Many of the mutations affecting β-myosin heavy chain involve the head and head–rod junction of the heavy chain (Fig. 12.3); some of these lead to pathologic changes early in life and produce severe hypertrophy. The clinical course varies even among persons with these mutations; an arginine-to-glutamine substitution at position 403 (Arg403Gln) and an arginine-to-tryptophan substitution at position 719 (Arg719Trp) predispose persons to sudden death and heart failure, whereas a phenylalanine-to-cysteine substitution at position 513 (Phe513Cys), a leucine-to-valine substitution at position 908 (Leu908Val), and a glycine-to-glutamic acid substitution at position 256 (Gly256Glu) cause less severe clinical disease.[30,36] In contrast, mutations affecting cardiac myosin-binding protein produce late-onset hypertrophic cardiomyopathy and are associated with a more favorable prognosis.[37]

Numerous factors other than sarcomere mutations determine the pathologic features and clinical course of hypertrophic cardiomyopathy. The identical sarcomere mutation can cause different hypertrophic changes and clinical outcomes among kindreds, even within the same pedigree.[38,39] Gene modifiers, the environment, sex, and acquired conditions (such as ischemic or valvular heart disease) may account for these differences. Studies examining polymorphisms in the genes encoding angiotensin II, aldosterone, and endothelin that may modify the phenotype of hypertrophic cardiomyopathy have not yielded consistent results.[40–43] Interestingly, clinically affected persons with two mutations in the same gene or different genes (compound heterozygotes) have also been described.[44]

Cardiac Arrhythmias

In 2001, about 450,000 people in the United States died suddenly from cardiac arrhythmias.[1,2] Genetic factors may modify the risk of arrhythmia in the setting of common environmental risks. Arrhythmia-susceptibility genes have been identified and provide insight into the molecular pathogenesis of lethal and nonlethal arrhythmias (Table 12.3).[45] The *SCN5A* gene encodes α subunits that form the sodium channels responsible for initiating cardiac action potentials.[46] Mutations in *SCN5A* cause several familial forms of arrhythmias, including the long-QT syndrome, idiopathic ventricular fibrillation, and cardiac-conduction disease.[47–51] A recently identified variant of the *SCN5A* gene, one with a transversion of cytosine to adenine in codon 1102, causing a serine-to-tyrosine substitution at position 1102 (S1102Y), has been associated with arrhythmia in black Americans.[52] The variant allele (Y1102) accelerates sodium-channel activation and increases the likelihood of abnormal cardiac repolarization and arrhythmia. About 13 percent of blacks carry one Y1102 allele,[52] which does not cause arrhythmia in most carriers. However, studies such

Table 12.3 Monogenic Diseases That Cause Susceptibility to Arrhythmias*

Disease	Mutant Gene	Mutations	Molecular Mechanism	Clinical Effect
Long-QT syndrome	*SCN5A*	Gain of function	Activation of the mutant sodium channel is normal, but channels reopen during plateau phase of the action potential	Repolarization abnormality
	KVLQT1 (KCNQ1)	Missense	minK β subunits assemble with KVLQT1 α subunits to form I_{Ks} potassium channels; mutations cause improper assembly of channels with reduced function	Repolarization abnormality
	minK (KCNE1)	Loss of function	Homozygous mutations cause Jervell and Lange–Nielson syndrome from loss of functional I_{Ks} potassium channels in the inner ear, leading to deafness	Repolarization abnormality
	HERG (KCNH2) and *MiRP-1 (KCNE2)*		HERG α subunits assemble with β subunits of minK-related peptide 1 (MiRP-1) to form I_{Kr} potassium channels; loss of function in HERG mutations in the membrane-spanning domain and pure region lead to a dominant negative suppression of channel function	Repolarization abnormality
Idiopathic ventricular fibrillation†	*SCN5A*	Loss of function	Reduction in the total number of functional sodium channels and expression of a heterogeneous group of sodium channels may shorten action potentials and slow conduction velocity	Conduction abnormality
Catecholamine-induced ventricular tachycardia	*RyR2*	Missense	Stress-induced calcium overload in myocytes may be a mechanism	Calcium overload in myocytes, leading to ventricular tachycardia

*The information is adapted from Keating and Sanguinetti,[45] with the permission of the publisher.

† Idiopathic ventricular fibrillation is also known as familial ventricular fibrillation.

as this point to the usefulness of molecular markers for the prediction of susceptibility to arrhythmia in persons with acquired or other genetic risk factors.

The *HERG* gene encodes α subunits that assemble with β subunits of minK-related peptide 1 (MiRP-1) to form cardiac I_{Kr} potassium channels, which facilitate a repolarizing potassium current.[53,54] In turn, *KVLQT1* α subunits assemble with minK β subunits to form cardiac I_{Ks} potassium channels, which facilitate a second repolarizing potassium current.[55,56] These channels terminate the plateau phase of the action potential, causing myocyte repolarization. *KVLQT1, HERG, minK,* and *MiRP-1* mutations result in a loss of function in the potassium channel that leads to the long-QT syndrome by reducing the repolarizing current. *RyR2* encodes the ryanodine-receptor calcium-release channel required for excitation–contraction coupling. Gain-of-function mutations in *SCN5A* cause the long-QT syndrome, whereas loss-of-function mutations in the cardiac sodium channel cause idiopathic ventricular fibrillation. *RyR2* mutations cause catecholaminergic ventricular tachycardia. Thus, inherited arrhythmia-susceptibility genes encode cardiac ion channels. Polymorphisms associated with inherited forms of the long-QT syndrome also increase the risk of acquired arrhythmias, such as drug-induced arrhythmias.[57]

ANALYSIS OF COMPLEX CARDIOVASCULAR TRAITS

Although many single genes have been identified as the basis of monogenic cardiovascular disorders, fewer genes underlying common complex cardiovascular diseases have been identified.[58] Multiple risk factors, gene–environment interactions, and an absence of rough estimates of the number of genes that influence a single trait all complicate study design. Current research on complex cardiovascular traits focuses on the identification of genetic variants that enhance the susceptibility to given conditions.

Gene Polymorphisms

Association studies provide a powerful approach to identifying DNA variants underlying complex cardiovascular traits and are very useful for narrowing a candidate interval identified by linkage analysis. Improved genotyping techniques, such as genome-wide scanning of single-nucleotide polymorphisms[59–61] and mapping of single-nucleotide polymorphisms identifying common haplotypes in the human genome, are facilitating association studies of loci spanning the entire genome. This point can be illustrated by recent examples of case–control studies that used high-throughput genomic techniques to investigate genetic variants in a large number of candidate genes for myocardial infarction, premature coronary artery disease, and heart failure.

Polymorphism-association studies compare the prevalence of a genetic marker in unrelated people with a given disease to the prevalence in a control population. Polymorphism-association studies of cardiovascular disease should be interpreted with caution when biologic plausibility has not been determined or is not known. Single-nucleotide polymorphisms in linkage disequilibrium may be functionally important, or alternatively, the polymorphism may just be a marker for another, yet to be identified, disease-causing sequence variant.

To determine genetic variants in myocardial infarction, Yamada and colleagues examined the prevalence of 112 polymorphisms in 71 candidate genes in patients with myocardial infarction and control patients in Japan.[62] The analysis revealed one statistically significant association in men (a cytosine-to-thymine polymorphism at nucleotide 1019 in the connexin 37 gene) and two in women (the replacement of four guanines with five guanines at position –668 [4G–668/5G] in the plasminogen-activator inhibitor type 1 gene and the replacement of five adenines with six adenines at position –1171 [5A–1171/6A] in the stromelysin-1 gene), suggesting that these single-nucleotide polymorphisms may confer susceptibility to myocardial infarction in this population.

The GeneQuest study investigated 62 candidate genes in patients and their siblings with premature myocardial infarction (men <45 years old and women <50 years old).[63] In this study, a case–control approach comparing genomic sequences in 72 single-nucleotide polymorphisms between persons with premature coronary artery disease and members of a control population identified three variants in the genes encoding thrombospondin-4, thrombospondin-2, and thrombospondin-1 that showed a statistical association with premature coronary artery disease. The biologic mechanisms by which these variants in thrombospondin proteins may lead to early myocardial infarction have yet to be identified.

Small and colleagues described an association between two polymorphisms in adrenergic-receptor genes and the risk of congestive heart failure in black Americans.[64] Genotyping at two loci—one encoding a variant α_{2c}-adrenergic receptor (involving the deletion of four amino acids [α_{2c}Del322–325]) and the other encoding a variant β_1-adrenergic receptor (with a glycine at amino acid position 389 [β_1Arg389])—was performed in patients with heart failure and in controls. The α_{2c}Del322–325 variant, when present alone, conferred some degree of risk, whereas the β_1Arg389 variant alone did not. However, black patients who were homozygous for both variants had a markedly increased incidence of heart failure. The presence of the α_{2c}Del322–325 variant is associated with norepinephrine release at cardiac sympathetic-nerve synapses, and the presence of the β_1Arg389 variant may increase the sensitivity of cardiomyocytes to norepinephrine. The findings of this study suggest that the α_{2c}Del322–325 and β_1Arg389 receptors act synergistically in blacks to increase the risk of heart failure. Genotyping at these two loci may identify persons at risk for the development or progression of heart failure and may predict their response to therapy.

These studies highlight the importance of cardiovascular genotyping to establish a molecular diagnosis, to stratify patients according to risk, and especially to guide therapy. The field of pharmacogenetics—that is, the use of genome-wide approaches to determine the role of genetic variants in individual responses to drugs—has provided data showing that genetic polymorphisms of proteins involved in drug metabolism, transporters, and targets have important effects on the efficacy of cardiovascular drugs[65,66] (Table 12.4). For example, sequence variants in the *ADRB2* gene, which encodes the β_2-adrenergic receptor, influence the response to β_2-agonist drugs.[79,87] Two common polymorphisms of the receptor, an arginine-to-glycine substitution at codon 16 (Gly16) and a glycine-to-glutamine substitution at codon 27 (Glu27), are associated with increased

Table 12.4 Examples of Polymorphisms in Genes That Alter Cardiovascular Function and Responses to Drugs

Gene*	Drug	Effect of Polymorphism on Response to Drug	Selected References
ABCB1 (MDR1)	Digoxin	Increased bioavailability, atrial arrhythmias, and heart failure	Hoffmeyer et al.,[67] Sakaeda et al.[68]
ACE	Angiotensin-converting–enzyme inhibitors†	Decreased blood pressure	Stavroulakis et al.,[69] O'Toole et al.[70]
		Reduction in left ventricular mass	Myerson et al.,[71] Kohno et al.[72]
		Survival after cardiac transplantation	McNamara et al.[73]
		Improvement in endothelium-dependent vasodilation	Prasad et al.[74]
		Renal protection	Yoshida et al.,[75] Perna et al.,[76] Penno et al.[77]
	Statins	Decreased LDL levels and regression of atherosclerosis	Marian et al.[78]
ADRB2	β_2-Adrenergic agonists	Vasodilation and bronchodilation	Dishy et al.,[79] Drysdale et al.,[80] Cockcroft et al.[81]
APOE	Statins	Decreased LDL levels and reduced mortality after myocardial infarction	Gerdes et al.,[82] Pedro-Botet et al.,[83] Ballantyne et al.[84]
CETP		Progression of coronary-artery atherosclerosis	Kuivenhoven et al.[85]
KCNE2 (MiRP-1)	Clarithromycin	Long-QT syndrome and ventricular fibrillation	Abbott et al.[54]
	Sulfamethoxazole	Long-QT syndrome	Sesti et al.[86]

**ACE* encodes angiotensin-converting enzyme, *ADRB2* encodes β_2-adrenergic receptor, *APOE* encodes apolipoprotein E, *CETP* encodes cholesterol ester transport protein, and *KCNE2* encodes minK-related peptide (MiRP-1). LDL denotes low-density lipoprotein.
†The response to angiotensin-converting–enzyme inhibitors is most pronounced in persons who are homozygous for a deletion in intron 16 of the *ACE* gene (i.e., those with the *D/D* genotype).

agonist-induced desensitization and increased resistance to desensitization, respectively. There is marked linkage disequilibrium between the polymorphisms at codons 16 and 27, with the result that persons who are homozygous for Glu27 are also likely to be homozygous for Gly16, whereas those who are homozygous for Gly16 may be homozygous for Gln27 or Glu27 or heterozygous at codon 27.

In a study that examined the effects of agonist-induced desensitization in the vasculature mediated by these polymorphisms, the investigators found that persons who were homozygous for Arg16 had nearly complete desensitization, as determined by measures of venodilation in response to isoproterenol, in contrast to persons homozygous for Gly16 and regardless of the codon 27 status.[79] Similarly, persons homozygous for Gln27 had higher maximal venodilation in response to isoproterenol than those homozygous for Glu27, regardless of their codon 16 status. These data demonstrate that polymorphisms of the β_2-adrenergic receptor are important determinants of vascular function. This study also highlights the importance of taking into account haplotypes, rather than a single polymorphism, when defining biologic function.

Gene-Expression Profiling

Functional genomics, which is the study of gene function by means of parallel measurements of expression within control and experimental genomes, commonly involves the use of microarrays and serial analysis of gene expression. Microarrays are artificially constructed grids of DNA in which each element of the grid acts as a probe for a specific RNA sequence; each grid holds a DNA sequence that is a reverse complement to the target RNA sequence. Measurements of gene expression by means of microarrays are useful tools to establish molecular diagnoses, dissect the pathophysiologic features of a disease, and predict patients' response to therapy.[88] Microarray analyses have been used to define a role for proliferative and inflammatory genes in the development of restenosis after the placement of coronary-artery stents. Zohlnhöfer and colleagues identified clusters of differentially expressed genes from coronary-artery neointima and peripheral-blood cells from patients with restenosis, as compared with samples of normal coronary arteries.[89,90] The up-regulation of genes with functions in cell proliferation, the synthesis of extracellular matrix, cell adhesion, and inflammatory responses were more abundant in the samples of neointima from the patients with restenosis than in the samples of normal coronary artery. Many genes were expressed to a similar extent in the neointimal tissues and the peripheral-blood cells from patients with restenosis, suggesting the possible use of peripheral-blood cells as a substitute in microarray studies when cardiovascular tissue is not available.

CONSIDERATIONS FOR MOLECULAR AND CLINICAL DIAGNOSIS

Genetic diagnosis—that is, primary classification on the basis of the presence of a mutation, with subsequent stratification according to risk—is not widely available for the diagnosis of monogenic cardiovascular disorders. Today, physical examination and routine testing, such as echocardiography to detect hypertrophic cardiomyopathy or electrocardiographic analysis of the long-QT syndrome, establish clinical diagnoses.[91] Genetic diagnoses are then made by research-oriented genotyping of selected pedigrees. Current initiatives focus on the natural history of monogenic disorders in large numbers of patients with specific mutations, in order to identify persons at high risk for cardiovascular events, asymptomatic carriers in whom pharmacologic interventions will retard or prevent disease, and nonaffected family members whose concern about their health can be addressed. With regard to complex traits in more common cardiovascular diseases, current research is identifying functionally significant variations in DNA sequences that can establish a molecular diagnosis and influence patients' outcome.

Acknowledgments

I am indebted to the members of my laboratory for their helpful comments.

References

1. NHLBI morbidity and mortality chartbook, 2002. Bethesda, Md.: National Heart, Lung, and Blood Institute, May 2002. (Accessed June 10, 2003, at http://www.nhlbi.nih.gov/resources/docs/cht-book.htm.)

2. NHLBI fact book, fiscal year 2002. Bethesda, Md.: National Heart, Lung, and Blood Institute, February 2003. (Accessed June 10, 2003, at http://www.nhlbi.nih.gov/about/factpdf.htm.)

3. Cooper R, Cutler J, Desvignes-Nickens P, et al. Trends and disparities in coronary heart disease, stroke, and other cardiovascular diseases in the United States: findings of the National Conference on Cardiovascular Disease Prevention. Circulation 2000; 102:3137–47.

4. Goldstein JL, Brown MS. The cholesterol quartet. Science 2001; 292:1310–2.

5. Goldstein JL, Hobbs HH, Brown MS. Familial hypercholesterolemia. In: Scriver CR, Beaudet AL, Sly WS, Valle D, eds. The metabolic & molecular bases of inherited disease. 8th ed. Vol. 2. New York: McGraw-Hill, 2001:2863–913.

6. Kane JP, Havel RJ. Disorders of the biogenesis and secretion of lipoproteins containing the B apolipoproteins. In: Scriver CR, Beaudet AL, Sly WS, Valle D, eds. The metabolic & molecular bases of inherited disease. 8th ed. Vol. 2. New York: McGraw-Hill, 2001:2717–52.

7. Berge KE, Tian H, Graf GA, et al. Accumulation of dietary cholesterol in sitosterolemia caused by mutations in adjacent ABC transporters. Science 2000; 290:1771–5.

8. Lee M-H, Lu K, Hazard S, et al. Identification of a gene, ABCG5, important in the regulation of dietary cholesterol absorption. Nat Genet 2001; 27:79–83.

9. Garcia CK, Wilund K, Arca M, et al. Autosomal recessive hypercholesterolemia caused by mutations in a putative LDL receptor adaptor protein. Science 2001; 292:1394–8.

10. ALLHAT Officers and Coordinators for the ALLHAT Collaborative Research Group. Major outcomes in high-risk hypertensive patients randomized to angiotensin-converting enzyme inhibitor or calcium channel blocker vs. diuretic: the Hypertensive and Lipid-Lowering Treatment to Prevent Heart Attack (ALLHAT). JAMA 2002; 288:2981–97. [Erratum, JAMA 2003; 289:178.]

11. Lifton RP, Gharavi AG, Geller DS. Molecular mechanisms of human hypertension. Cell 2001; 104:545–56.

12. Wilson FH, Disse-Nicodeme S, Choate KA, et al. Human hypertension caused by mutations in WNK kinases. Science 2001; 293:1107–12.

13. Lifton RP, Dluhy RG, Powers M, et al. A chimaeric 11 beta-hydroxylase/aldosterone synthase gene causes glucocorticoid-remediable aldosteronism and human hypertension. Nature 1992; 355:262–5.

14. Lifton RP, Dluhy RG, Powers M, et al. Hereditary hypertension caused by chimaeric gene duplications and ectopic expression of aldosterone synthase. Nat Genet 1992; 2:66–74.

15. Mitsuuchi Y, Kawamoto T, Naiki Y, et al. Congenitally defective aldosterone biosynthesis in humans: the involvement of point mutations of the P-450C18 gene (CYP11B2) in CMO II deficient patients. Biochem Biophys Res Commun 1992; 182:974–9. [Erratum, Biochem Biophys Res Commun 1992; 184:1529–30.]

16. Shimkets RA, Warnock DG, Bositis CM, et al. Liddle's syndrome: heritable human hypertension caused by mutations in the beta subunit of the epithelial sodium channel. Cell 1994; 79:407–14.

17. Hansson JH, Schild L, Lu Y, et al. A *de novo* missense mutation of the beta subunit of the epithelial sodium channel causes hypertension and Liddle syndrome, identifying a proline-rich segment critical for regulation of channel activity. Proc Natl Acad Sci U S A 1995; 92:11495–9.

18. Simon DB, Nelson-Williams C, Bia MJ, et al. Gitelman's variant of Bartter's syndrome, inherited hypokalaemic alkalosis, is caused by mutations in the thiazide-sensitive Na-Cl cotransporter. Nat Genet 1996; 12:24–30.

19. Simon DB, Karet FE, Hamdan JM, DiPietro A, Sanjad SA, Lifton RP. Bartter's syndrome, hypokalaemic alkalosis with hypercalciuria, is caused by mutations in the Na-K-2Cl cotransporter NKCC2. Nat Genet 1996; 13:183–8.

20. Svensson PJ, Dahlbäck B. Resistance to activated protein C as a basis for venous thrombosis. N Engl J Med 1994; 330:517–22.

21. Griffin JH, Evatt B, Wideman C, Fernández JA. Anticoagulant protein C pathway defective in majority of thrombophilic patients. Blood 1993; 82:1989–93.

22. Koster T, Rosendaal FR, de Ronde H, Briët E, Vandenbroucke JP, Bertina RM. Venous thrombosis due to poor anticoagulant response to activated protein C: Leiden Thrombophilia Study. Lancet 1993; 342:1503–6.

23. Greengard JS, Eichinger S, Griffin JH, Bauer KA. Variability of thrombosis among homozygous siblings with resistance to activated protein C due to an Arg → Gln mutation in the gene for factor V. N Engl J Med 1994; 331:1559–62.

24. Rosendaal FR, Koster T, Vandenbroucke JP, Reitsma PH. High risk of thrombosis in patients homozygous for factor V Leiden (activated protein C resistance). Blood 1995; 85:1504–8.

25. Ridker PM, Hennekens CH, Lindpaintner K, Stampfer MJ, Eisenberg PR, Miletich JP. Mutation in the gene coding for coagulation factor V and the risk of myocardial infarction, stroke, and venous thrombosis in apparently healthy men. N Engl J Med 1995; 332:912–7.

26. Koeleman BPC, Reitsma PH, Allaart CF, Bertina RM. Activated protein C resistance as an additional risk factor for the thrombosis in protein C-deficient families. Blood 1994; 84:1031–5.

27. Zöller B, Berntsdotter A, Garcia de Frutos P, Dahlbäck B. Resistance to activated protein C as an additional genetic risk factor in hereditary deficiency of protein S. Blood 1995; 85:3518–23.

28. Van Boven HH, Reitsma PH, Rosendaal FR, et al. Factor V Leiden (FV R506Q) in families with inherited antithrombin deficiency. Thromb Haemost 1996; 75:417–21.

29. De Stefano V, Martinelli I, Mannucci PM, et al. The risk of recurrent venous thromboembolism among heterozygous carriers of the G20210A prothrombin gene mutation. Br J Haematol 2001; 113:630–5.

30. Seidman CE, Seidman JG. Hypertrophic cardiomyopathy. In: Scriver CR, Beaudet AL, Sly WS, Valle D, eds. The metabolic & molecular bases of inherited disease. 8th ed. Vol. 4. New York: McGraw-Hill, 2001:5433–50.

31. Maron BJ, Gardin JM, Flack JM, Gidding SS, Kurosaki TT, Bild DE. Prevalence of hypertrophic cardiomyopathy in a general population of young adults: echocardiographic analysis of 4111 subjects in the CARDIA Study: Coronary Artery Risk Development in (Young) Adults. Circulation 1995; 92:785–9.

32. Kamisago M, Sharma SD, DePalma SR, et al. Mutations in sarcomere protein genes as a cause of dilated cardiomyopathy. N Engl J Med 2000; 343:1688–96.

33. Seidman JG, Seidman C. The genetic basis for cardiomyopathy: from mutation identification to mechanistic paradigms. Cell 2001; 104:557–67.

34. Geisterfer-Lowrance AA, Kass S, Tanigawa G, et al. A molecular basis for familial hypertrophic cardiomyopathy: a beta cardiac myosin heavy chain gene missense mutation. Cell 1990; 62:999–1006.

35. NHLBI Program for Genomic Applications. Genomics of cardiovascular development, adaptation, and remodeling. Boston: Beth Israel Deaconess Medical Center, 2003. (Accessed June 10, 2003, at http://cardiogenomics.med.harvard.edu/project-detail?project_id = 230#data.)

36. Roberts R, Sigwart U. New concepts in hypertrophic cardiomyopathy. Circulation 2001; 104:2113–6.

37. Niimura H, Bachinski LL, Sangwatanaroj S, et al. Mutations in the gene for cardiac myosin-binding protein C and late-onset familial hypertrophic cardiomyopathy. N Engl J Med 1998; 338:1248–57.

38. Fananapazir L, Epstein N. Genotype-phenotype correlations in hypertrophic cardiomyopathy: insights provided by comparisons of kindreds with distinct and identical β-myosin heavy chain gene mutations. Circulation 1994; 89:22–32.

39. Kimura A, Harada H, Park JE, et al. Mutations in the cardiac troponin I gene associated with hypertrophic cardiomyopathy. Nat Genet 1997; 16:379–82.

40. Lechin M, Quinones MA, Omran A, et al. Angiotensin-I converting enzyme genotypes and left ventricular hypertrophy in patients with hypertrophic cardiomyopathy. Circulation 1995; 92:1808–12.

41. Tesson F, Dufour C, Moolman JC, et al. The influence of the angiotensin I converting enzyme genotype in familial hypertrophic cardiomyopathy varies with the disease gene mutation. J Mol Cell Cardiol 1997; 29:831–8.

42. Osterop AP, Kofflard MJ, Sandkuijl LA, et al. AT1 receptor A/C1166 polymorphism contributes to cardiac hypertrophy in subjects with hypertrophic cardiomyopathy. Hypertension 1998; 32:825–30.

43. Ortlepp JR, Vosberg HP, Reith S, et al. Genetic polymorphisms in the renin-angiotensin-aldosterone system associated with expression of left

ventricular hypertrophy in hypertrophic cardiomyopathy: a study of five polymorphic genes in a family with a disease causing mutation in the myosin binding protein C gene. Heart 2002; 87:270–5.

44. Ho CY, Lever HM, DeSanctis R, Farver CF, Seidman JG, Seidman CE. Homozygous mutation in cardiac troponin T: implications for hypertrophic cardiomyopathy. Circulation 2000; 102:1950–5.

45. Keating MT, Sanguinetti MC. Molecular and cellular mechanisms of cardiac arrhythmias. Cell 2001; 104:569–80.

46. Gellens ME, George AL Jr, Chen LQ, et al. Primary structure and functional expression of the human cardiac tetrodotoxin-insensitive voltage-dependent sodium channel. Proc Natl Acad Sci U S A 1992; 89:554–8.

47. Wang Q, Shen J, Splawski I, et al. SCN5A mutations associated with an inherited cardiac arrhythmia, long QT syndrome. Cell 1995; 80: 805–11.

48. Bennett PB, Yazawa K, Makita N, George AL Jr. Molecular mechanism for an inherited cardiac arrhythmia. Nature 1995; 376:683–5.

49. Chen Q, Kirsch GE, Zhang D, et al. Genetic basis and molecular mechanism for idiopathic ventricular fibrillation. Nature 1998; 392:293–6.

50. Schott JJ, Alshinawi C, Kyndt F, et al. Cardiac conduction defects associate with mutations in SCN5A. Nat Genet 1999; 23:20–1.

51. Tan HL, Bink-Boelkens MT, Bezzina CR, et al. A sodium-channel mutation causes isolated cardiac conduction disease. Nature 2001; 409:1043–7.

52. Splawski I, Timothy KW, Tateyama M, et al. Variant of SCN5A sodium channel implicated in risk of cardiac arrhythmia. Science 2002; 297:1333–6.

53. Sanguinetti MC, Jiang C, Curran ME, Keating MT. A mechanistic link between an inherited and an acquired cardiac arrhythmia: HERG encodes the I_{Kr} potassium channel. Cell 1995; 81:299–307.

54. Abbott GW, Sesti F, Splawski I, et al. MiRP1 forms I_{Kr} potassium channels with HERG and is associated with cardiac arrhythmia. Cell 1999; 97:175–87.

55. Barhanin J, Lesage F, Guillemare E, Fink M, Lazdunski M, Romey G. K_VLQT1 and IsK (minK) proteins associate to form the I_{Ks} cardiac potassium current. Nature 1996; 384:78–80.

56. Sanguinetti MC, Curran ME, Zou A, et al. Coassembly of K_VLQT1 and minK (IsK) proteins to form cardiac I_{Ks} potassium channel. Nature 1996; 384:80–3.

57. Splawski I, Shen J, Timothy KW, et al. Spectrum of mutations in long-QT syndrome genes: KVLQT1, HERG, SCN5A, KCNE1, and KCNE2. Circulation 2000; 102:1178–85.

58. Glazier AM, Nadeau JH, Aitman TJ. Finding genes that underlie complex traits. Science 2002; 298:2345–9.

59. Mohlke KL, Erdos MR, Scott LJ, et al. High-throughput screening for evidence of association by using mass spectrometry genotyping on DNA pools. Proc Natl Acad Sci U S A 2002; 99:16928–33.

60. Oliphant A, Barker DL, Stuelpnagel JR, Chee MS. BeadArray technology: enabling an accurate, cost-effective approach to high-throughput genotyping. Biotechniques 2002; Suppl:56–8, 60–1.

61. Cargill M, Altshuler D, Ireland J, et al. Characterization of single-nucleotide polymorphisms in coding regions of human genes. Nat Genet 1999; 22:231–8. [Erratum, Nat Genet 1999; 23:373.]

62. Yamada Y, Izawa H, Ichihara S, et al. Prediction of the risk of myocardial infarction from polymorphisms in candidate genes. N Engl J Med 2002; 347:1916–23.

63. Topol EJ, McCarthy J, Gabriel S, et al. Single nucleotide polymorphisms in multiple novel thrombospondin genes may be associated with familial premature myocardial infarction. Circulation 2001; 104:2641–4.

64. Small KM, Wagoner LE, Levin AM, Kardia SLR, Liggett SB. Synergistic polymorphisms of β_1- and α_{2c}-adrenergic receptors and the risk of congestive heart failure. N Engl J Med 2002; 347:1135–42.

65. Weinshilboum R. Inheritance and drug response. N Engl J Med 2003; 348:529–37.

66. Evans WE, McLeod HL. Pharmacogenomics—drug disposition, drug targets, and side effects. N Engl J Med 2003; 348:538–49.

67. Hoffmeyer S, Burk O, von Richter O, et al. Functional polymorphisms of the human multidrug-resistance gene: multiple sequence variations and correlation of one allele with P-glycoprotein expression and activity in vivo. Proc Natl Acad Sci U S A 2000; 97:3473–8.

68. Sakaeda T, Nakamura T, Horinouchi M, et al. MDR1 genotype-related pharmacokinetics of digoxin after single oral administration in healthy Japanese subjects. Pharm Res 2001; 18:1400–4.

69. Stavroulakis GA, Makris TK, Krespi PG, et al. Predicting response to chronic antihypertensive treatment with fosinopril: the role of angiotensin-converting enzyme gene polymorphism. Cardiovasc Drugs Ther 2000; 14:427–32.

70. O'Toole L, Stewart M, Padfield P, Channer K. Effect of the insertion/deletion polymorphism of the angiotensin-converting enzyme gene on response to angiotensin-converting enzyme inhibitors in patients with heart failure. J Cardiovasc Pharmacol 1998; 32:988–94.

71. Myerson SG, Montgomery HE, Whittingham M, et al. Left ventricular hypertrophy with exercise and ACE gene insertion/deletion polymorphism: a randomized controlled trial with losartan. Circulation 2001; 103:226–30.

72. Kohno M, Yokokawa K, Minami M, et al. Association between angiotensin-converting enzyme gene polymorphisms and regression of left ventricular hypertrophy in patients treated with angiotensin-converting enzyme inhibitors. Am J Med 1999; 106:544–9.

73. McNamara DM, Holubkov R, Janosko K, et al. Pharmacogenetic interactions between β-blocker therapy and the angiotensin-converting enzyme deletion polymorphism in patients with congestive heart failure. Circulation 2001; 103:1644–8.

74. Prasad A, Narayanan S, Husain S, et al. Insertion-deletion polymorphism of the ACE gene modulates reversibility of endothelial dysfunction with ACE inhibition. Circulation 2000; 102:35–41.

75. Yoshida H, Mitarai T, Kawamura T, et al. Role of the deletion of polymorphism of the angiotensin converting enzyme gene in the progression and therapeutic responsiveness of IgA nephropathy. J Clin Invest 1995; 96:2162–9.

76. Perna A, Ruggenenti P, Testa A, et al. ACE genotype and ACE inhibitors induced renoprotection in chronic proteinuric nephropathies. Kidney Int 2000; 57:274–81.

77. Penno G, Chaturvedi N, Talmud PJ, et al. Effect of angiotensin-converting enzyme (ACE) gene polymorphism on progression of renal disease and the influence of ACE inhibition in IDDM patients: findings from the EUCLID Randomized Controlled Trial: EURODIAB Controlled Trial of Lisinopril in IDDM. Diabetes 1998; 47:1507–11.

78. Marian AJ, Safavi F, Ferlic L, Dunn JK, Gotto AM, Ballantyne CM. Interactions between angiotensin-I converting enzyme insertion/deletion polymorphism and response of plasma lipids and coronary atherosclerosis to treatment with fluvastatin: the Lipoprotein and Coronary Atherosclerosis Study. J Am Coll Cardiol 2000; 35:89–95.

79. Dishy V, Sofowora GG, Xie H-G, et al. The effect of common polymorphisms of the β_2-adrenergic receptor on agonist-mediated vascular desensitization. N Engl J Med 2001; 345:1030–5.

80. Drysdale CM, McGraw DW, Stack CB, et al. Complex promoter and coding region beta 2-adrenergic receptor haplotypes alter receptor expression and predict in vivo responsiveness. Proc Natl Acad Sci U S A 2000; 97:10483–8.

81. Cockcroft JR, Gazis AG, Cross DJ, et al. Beta-2 adrenoreceptor polymorphism determines vascular reactivity in humans. Hypertension 2000; 36:371–5.

82. Gerdes LU, Gerdes C, Kervinen K, et al. The apolipoprotein epsilon4 allele determines prognosis and the effect on prognosis of simvastatin in survivors of myocardial infarction: a substudy of the Scandinavian Simvastatin Survival Study. Circulation 2000; 101:1366–71.

83. Pedro-Botet J, Schaefer EJ, Bakker-Arkema RG, et al. Apolipoprotein E genotype affects plasma lipid response to atorvastatin in a gender specific manner. Atherosclerosis 2001; 158:183–93.

84. Ballantyne CM, Herd JA, Stein EA, et al. Apolipoprotein E genotypes and response of plasma lipids and progression-regression of coronary atherosclerosis to lipid-lowering drug therapy. J Am Coll Cardiol 2000; 36:1572–8.

85. Kuivenhoven JA, Jukema JW, Zwinderman AH, et al. The role of a common variant of the cholesteryl ester transfer protein gene in the progression of coronary atherosclerosis. N Engl J Med 1998; 338:86–93.

86. Sesti F, Abbott GW, Wei J, et al. A common polymorphism associated with antibiotic-induced cardiac arrhythmia. Proc Natl Acad Sci U S A 2000; 97:10613–8.

87. Liggett SB. Beta(2)-adrenergic receptor pharmacogenetics. Am J Respir Crit Care Med 2000; 161:S197–S201.

88. Rosenwald A, Wright G, Wiestner A, et al. The proliferation gene expression signature is a quantitative integrator of oncogenic events that predicts survival in mantle cell lymphoma. Cancer Cell 2003; 3:185–97.

89. Zohlnhöfer D, Richter T, Neumann F-J, et al. Transcriptome analysis reveals a role of interferon-γ in human neointima formation. Mol Cell 2001; 7:1059–69.

90. Zohlnhöfer D, Klein CA, Richter T, et al. Gene expression profiling of human stent-induced neointima by cDNA array analysis of microscopic specimens retrieved by helix cutter atherectomy: detection of FK506-binding protein 12 upregulation. Circulation 2001; 103:1396–402.

91. Maron BJ, Moller JH, Seidman CE, et al. Impact of laboratory molecular diagnosis on contemporary diagnostic criteria for genetically transmitted cardiovascular diseases: hypertrophic cardiomyopathy, long-QT syndrome, and Marfan syndrome: a statement for healthcare professionals from the councils on clinical cardiology, cardiovascular disease in the young, and basic science, American Heart Association. Circulation 1998; 98:1460–71.

13

Ethical, Legal, and Social Implications of Genomic Medicine

ELLEN WRIGHT CLAYTON, M.D., J.D.

Genomics has contributed greatly to our understanding of the molecular basis of disease and, to a lesser but growing extent, to the development of effective interventions. Clinicians and society at large, however, are concerned about the effect genetic knowledge will have on the well-being of individual persons and groups. Much effort is being devoted to trying to anticipate, understand, and address the ethical, legal, social, and political implications of genetics and genomics.

The inquiry is complex. Understanding the social effects of genomics requires an analysis of the ways in which genetic information and a genetic approach to disease affect people individually, within their families and communities, and in their social and working lives. Genomics presents particular challenges with respect to clinicians' ethical and professional responsibilities, including the appropriate use of genomic information in the health care setting. In this chapter, I examine public concerns about genetic information and discuss a few recent cases in some depth to highlight a few of the dilemmas presented by genomics and emerging solutions.

WHAT GENETIC INFORMATION IS AND WHAT PEOPLE ARE WORRIED ABOUT

Genes affect virtually all human characteristics and diseases. These influences can be ascertained in individual patients through a review of the family history, physical examination, and the use of medical diagnostics. In some conditions, such as cystic fibrosis and sickle cell disease, the specific molecular mechanisms are largely understood, but in many, including such common chronic diseases as diabetes mellitus and hypertension,

Originally published August 7, 2003

the relevant genes—and there are often many—are only beginning to be identified. Given the variety of these effects and the limits of our knowledge, it is not surprising that the term *genetic information* is used in different ways at different times. Sometimes it is used to mean the influence of the entire genome, but more often it is used to refer to recognized, single-gene disorders or, even more narrowly, the results of DNA-based tests. These various meanings may make sense in context, but confusion can occur unless the speaker and listener are defining the term in the same way.

The most commonly expressed fear is that genetic information will be used in ways that could harm people—for example, to deny them access to health insurance, employment, education, and even loans. Part of that concern is fueled by the growing recognition that health information is not entirely private, despite people's expectations and desires to the contrary. In fact, both federal and state governments have been actively engaged in discussions about who ought to have access to health information and under what conditions.[1,2] This debate is informed appropriately by the recognition that limiting access to the medical record to the patient and the treating clinician is neither possible nor unequivocally desirable.[3]

People tend to see genetic information as more definitive and predictive than other types of data, in the sense that "you cannot change your genes" and that "genes tell all about your future." This notion of genetic determinism, however, includes an unwarranted sense of inevitability, because it reflects a fundamental failure to understand the nature of biologic systems. The DNA sequence is not the Book of Life. Human characteristics are the product of complex interactions over time between genes—both a person's own and those of other organisms—and the environment. Both germ-line and somatic cells undergo mutations, the latter being a primary way in which cancer develops. Moreover, a pathogenic mutation does not doom one to ill health; many diseases can be treated. As is true for so many conditions in medicine, clinicians have a variable but usually limited ability to predict when, how severely, and even whether a person with a genetic predisposition to a certain illness is going to become ill.

One might be tempted to conclude that the way to allay people's fears about genetics is simply to give them a more realistic understanding of the informative power of these tools. Given the optimistic predictions about genetics that pervade the media and public opinion today, that path is unlikely to succeed in the short term. A more promising approach to addressing the social implications of genetics requires us to consider both how genes are perceived in the real world and what is actually known about their function.

THE PROBLEM OF DISCRIMINATION

The question of whether genetic information should ever be used to affect one's access to health and other forms of insurance has been a dominant issue of public concern in the past decade. People cite fear of losing insurance as a major reason to avoid genetic testing.[4] Others argue that discrimination by insurance companies is not a problem, often pointing out that few of these cases, which are difficult for employees to win, have been filed.[5] Insurers assert

Table 13.1 Summary of Statutes Regarding Discrimination on the Basis of Genetic Information and the Privacy of Such Information*				
State or District	Health Insurance	Life Insurance	Employment	Confidentiality
Alabama	Yes, for cancer only†			
Alaska	Yes‡			
Arizona	Yes	Yes	Yes	Yes
Arkansas	Yes†‡		Yes	Yes
California	Yes§	Yes§	Yes	Yes
Colorado	Yes	Yes		Yes
Connecticut	Yes‡		Yes§	Yes
Delaware	Yes		Yes§	Yes
District of Columbia	Yes‡			
Florida	Yes‡			Yes
Georgia	Yes			Yes¶
Hawaii	Yes†‡§			
Idaho	Yes‡			
Illinois	Yes‡			Yes
Indiana	Yes†§			Yes
Iowa	Yes†‡		Yes†	
Kansas	Yes†§		Yes†	
Kentucky	Yes†‡			
Louisiana	Yes†		Yes	Yes
Maine	Yes‡	Yes	Yes†	
Maryland	Yes†	Yes	Yes†	
Massachusetts	Yes†	Yes	Yes	Yes
Michigan	Yes†		Yes†‖	
Minnesota	Yes†§		Yes†	
Mississippi				
Missouri	Yes†		Yes	Yes
Montana	Yes†‡¶	Yes		
Nebraska	Yes‡		Yes†‖	
Nevada	Yes†‡§		Yes†	Yes
New Hampshire	Yes†	Yes	Yes	Yes
New Jersey	Yes‡	Yes	Yes	Yes
New Mexico	Yes‡			Yes¶

(continued)

Table 13.1 (*continued*)

State or District	Health Insurance	Life Insurance	Employment	Confidentiality
New York	Yes		Yes	Yes
North Carolina	Yes		Yes	
North Dakota	Yes‡			Yes
Ohio	Yes†‡			
Oklahoma	Yes†‡		Yes†	Yes†
Oregon	Yes		Yes†	Yes
Pennsylvania				
Rhode Island	Yes		Yes	
South Carolina	Yes§			Yes
South Dakota	Yes†‡		Yes	Yes
Tennessee	Yes‡			
Texas	Yes**		Yes	Yes¶
Utah	Yes		Yes	Yes
Vermont	Yes†	Yes	Yes†	Yes
Virginia	Yes§		Yes	
Washington				
West Virginia	Yes‡			
Wisconsin	Yes†	Yes	Yes†‖	
Wyoming	Yes‡			

*Yes indicates that the state has enacted legislation concerning the use of genetic information in the indicated circumstance. This table was compiled in June 2003. Because these are areas of intense legislative activity, the laws change frequently. In addition, the laws vary far more widely from state to state than can be reflected in a table such as this. This table is not intended to be a legal opinion about the coverage of these laws. Readers are encouraged to consult the laws in their own states.

†Testing cannot be required.

‡According to the statute, genetic information cannot be considered to indicate a preexisting condition in the absence of symptoms.

§The statute specifically addresses illnesses in family members.

¶The statute contains exemptions about the use of information for certain research and other purposes.

‖Testing can be required for certain purposes, such as evaluating workers' compensation claims or surveillance.

**The statute permits testing to be required under certain circumstances.

that they do not perform tests to obtain genetic information but argue that they should be free to use such information if it is available, citing the need to avoid "moral hazard"—the risk that people who know they will become ill or die soon will try to obtain insurance at regular rates.[6] In response to consumer pressure, many states have passed laws in this area (Table 13.1).[7,8] In passing the Health Insurance Portability and Accountability Act (HIPAA),[9] Congress specifically banned certain uses of genetic information in determining insurance eligibility, but it placed no limits on rate setting.[10] Vigorous debate about optimal solutions is ongoing,[11] and bills have been introduced in every recent session of Congress.[12]

The complexity of the issues surrounding discrimination can be illustrated more generally by examining a case involving Burlington Northern Santa Fe Railroad (BNSF). Allegedly relying on the advice of its company physician, who in turn had apparently relied on the representations of a diagnostic company, BNSF began obtaining blood for DNA testing from employees who were seeking disability compensation as a result of carpal tunnel syndrome that occurred on the job. The employees were reportedly not told the purpose of the tests, which was to detect a mutation associated with hereditary neuropathy with liability to pressure palsies.[13] The company's motive for pursuing testing was never made clear, but it seems reasonable to suspect that BNSF would have tried to deny disability benefits to any employee who had such a mutation, arguing that the mutation, and not the job, caused the carpal tunnel syndrome. When the company's practice came to light, it was almost immediately stopped by the federal Equal Employment Opportunity Commission,[14] and shortly thereafter, the company settled claims brought by its employees for an undisclosed amount of money.[15]

What lessons can be learned here? One is that the company's effort to find mutations for hereditary neuropathy with liability to pressure palsies made little sense. This disorder is very rare, affecting about 3 to 10 persons per 100,000, and more important, although carpal tunnel syndrome can be a part of hereditary neuropathy with liability to pressure palsies, it has not been reported as the sole symptom. The injuries these employees sustained were not the result of an epidemic of hereditary neuropathy with liability to pressure palsies. Getting the biologic process correct is a critical step in making decisions about genetic testing.

Another important lesson is that identifying a genetic predisposition to carpal tunnel syndrome would not have been the end of the discussion in the eyes of the law. The company got in trouble because its practice violated numerous laws forbidding discrimination in the workplace. In particular, the Americans with Disabilities Act permits employers to require a medical evaluation only under clearly specified circumstances.[16] Testing employees after they were disabled without their informed consent clearly fell outside the bounds of this and other antidiscrimination laws.

The actions of BNSF led to widespread criticism and, not surprisingly, to calls to ban genetic discrimination in the workplace.[17] Although some states have enacted laws (Table 13.1), the need for federal action has grown as the Supreme Court has progressively narrowed the protection provided under the Americans with Disabilities Act.[18,19] The answer, however, is not simply to forbid employers to use genetic information or to require genetic testing.

The first step in developing an appropriate response is to determine how the use of genetic information fits within the broader framework of antidiscrimination laws, which were passed to create a certain kind of society, one in which people must be included regardless of race, sex, or disability, even at some cost to employers. Biology alone does not determine the social outcome. To use an analogy, an employer cannot exclude women from the workplace, even if he or she believes, with some justification, that women are more likely than men to take time off to care for family members. At the same time, employers are not required to bear unlimited costs to promote these social goals—the employee, male or female, who misses months of work at a time to care for sick relatives can still be fired.

A similar debate about social goals and the limits of our pursuit of them must occur with regard to genetic discrimination. The Equal Employment Opportunity Commission recently awarded damages to Terri Sergeant, who was fired from her job as an office manager for an insurance broker because she required extremely expensive medication to treat her at-worst mildly symptomatic alpha$_1$-antitrypsin deficiency.[20] A person's need for expensive health care is not sufficient reason to fire that person or to refuse to hire him or her in the first place. The fact that the costs may cause the employer to go under or to decide not to provide health insurance simply underlines the inherent weakness of employment-based health insurance.

At the same time, one can imagine a genetic condition that might affect a person's ability to perform a job in ways that could not be accommodated with reasonable efforts. Suppose a person with a recurrent and untreatable cardiac arrhythmia that leads to loss of consciousness, owing to an inherited ion-channel defect, is seeking employment as a long-distance truck driver. Because of the risk to third parties, such a person would not even be able to get a driver's license in many jurisdictions. The more difficult question—and the one posed particularly with respect to genetics—would arise if an asymptomatic person had a predisposing, but incompletely penetrant, mutation for the same disorder. Deciding what to do about such predispositions will require close attention both to the true, as opposed to the feared, likelihood that symptoms will develop and to the complex weighing of the interests of the individual, the employer, and society.

A similar calculus must be applied to every question regarding who can obtain and use genetic information to distinguish, or discriminate, among people in ways that affect their ability to obtain social goods, such as health insurance and education (Table 13.2).[21] If,

Table 13.2 Elements to Be Considered in Decisions about the Use of Genetic Information

What are some potential implications of genetic information?
- The patient may be more likely to require expensive therapy
- The patient may be more likely to be injured by certain types of exposure
- The patient may present a danger to others in the future

What principles need to be taken into account, recognizing that none is absolute?
- Protection of autonomy
- The public health
- The importance of inclusiveness
- Allocation of costs

Who decides whether the test will be done?
- The patient
- The patient's employer or another private third party
- The government

Who decides what to do with the results?
- The patient
- A private third party
- The government

as is likely, some uses are deemed to be appropriate, the challenge for clinicians will be to discuss with their patients the potential adverse social consequences of testing so that the patients can make informed choices about whether or not to proceed with testing.

THE CHALLENGE OF GENOMIC MEDICINE WITH RESPECT TO THE PHYSICIAN–PATIENT RELATIONSHIP

Consider the case of a man who died of colon cancer in the 1960s. When the same disease developed in his daughter approximately 25 years later, she obtained her father's pathology slides, discovered that he had had diffuse adenomatous polyposis coli, and sued the estate of her father's surgeon, alleging that the physician should have warned her about her 50 percent risk of having the disorder. An intermediate appellate court in New Jersey ruled that the physician had a duty to warn the daughter directly (she would have been a child at the time of her father's death), perhaps even over her father's objections.[22]

This is only one court's view in one case, but given how much attention it received, it is important to ask whether this was a good result. Two central tenets of Western medicine are that physicians should focus on the interests of their patients and that they should protect the confidentiality of their patients' medical information. Yet the tools of genomic medicine often reveal information about health risks faced not only by patients but also by their relatives. What should clinicians do? It seems clear that they should tell their patients about the risks faced by family members. The harder questions are whether physicians are ethically permitted to contact the relatives themselves, in contravention of traditional patient-centered norms, and whether they should be legally required to do so.

This issue must be viewed in the light of the fact that the duty to protect confidentiality is not absolute. Physicians are required to report numerous infectious diseases,[23] and they have been held liable for failing to warn people whom their patients have specifically threatened with violence.[24] The question then becomes more complex: are genetic risks sufficiently similar to these existing exceptions to the requirement of confidentiality that they warrant an exception as well? Over the years, numerous prominent advisory bodies have said no, opining that physicians should be permitted to breach confidentiality in order to warn third parties of genetic risks only as a last resort to avert serious harm.[25–27]

These learned opinions, however, are not the end of the matter, in part because they lack the force of law. In fact, as the case above illustrates, relatives have sued the primary patients' physicians for failing to warn them of their own genetic risks—and won limited victories, although none has been awarded monetary damages. The decisions in the colon-cancer case and a similar one in Florida[28] have been criticized for both their legal reasoning and their deviation from ethical guidelines, but they have not been overturned and, in the tradition of the common law, may be persuasive to other courts. Physicians who breach their patients' confidentiality and warn family members are not likely to incur substantial liability, even under HIPAA.[29] As a result, physicians might understandably conclude that warning relatives is the least risky option.

The existing directives are thus in conflict: "expert consensus," ethical analysis, and the HIPAA regulations argue for honoring confidentiality, whereas at least one legal opinion holds that physicians fail to warn a patient's relatives at their peril. Given the press of other business, legislators are not likely to resolve this conflict soon. In this setting, clinicians should inform their patients about the risks their relatives face, discuss the appropriateness of sharing this information and offer assistance, trusting—usually realistically—that patients will in turn tell their relatives who are at risk, and hope that the courts will get it right in the future.

GENOMIC MEDICINE AND PUBLIC HEALTH

When Sierra Creason underwent state-mandated newborn screening, she had abnormally low levels of both thyroxine and thyrotropin, findings consistent with the presence of congenital hypothyroidism. Her physician was not notified of these results, however, because the state had chosen not to divulge the actual values and, instead, to report as abnormal only results in which thyroxine levels were low and thyrotropin levels were high.[30] As a result, the diagnosis of congenital hypothyroidism and subsequent treatment were seriously delayed, resulting in permanent harm. When the child's family sued the state, however, the California Supreme Court ruled that the state program could not be held liable, in part to avoid diverting funds that would have been used for other state purposes. By contrast, had a private diagnostic laboratory given the same report, especially without providing the actual results, which would have enabled the child's physician to make an independent assessment, it almost certainly would have been held responsible.

Complex questions arise when the government requires testing and interventions. State-mandated screening of newborns for metabolic and genetic disorders was described by Khoury et al. in Chapter 4.[31] Governments undertake many activities to promote health—universal screening of newborns for phenylketonuria, for example, is generally considered a resounding success—but it is worth asking in each case whether there is sufficient justification to pursue mandatory as opposed to voluntary action or to place such activities in the public rather than the private sector. Requiring public health agencies to assume such responsibilities has advantages, such as more transparent accountability to the public and greater uniformity in access and results. Relying on public health entities in matters that directly affect the health of individual persons, however, entails certain risks as well. Physicians and patients count on receiving accurate and informative results regardless of whether a private or a public entity is doing the testing. Permitting state agencies to avoid financial responsibility when their actions harm patients like Sierra Creason is unjust and should raise questions about the wisdom of proposals that would dramatically expand newborn screening.

A public health analysis of genomics, of course, involves more than state-run testing. The broadest question is whether the public's health is actually improved by the knowledge derived. A major determinant is access to testing and to the medical interventions that may be warranted as a result. In our current multipayer system of health care, people will have widely differing levels of access to these forms of technology. One cannot assume that everyone will reap the benefits of this knowledge.

From a public health perspective, it might do to go one step further and ask whether people will actually use the test results to alter their behavior in ways that improve health. Some people whom testing identifies as predisposed to cancer subsequently decline to undergo surveillance or other interventions for psychological reasons or because of other demands on their time. Some preventive or therapeutic measures are more likely to be pursued than others; most people find it difficult to take medications for a lifetime or to maintain major lifestyle changes, no matter how important such approaches are for their health.

Public health agencies exist not only to identify barriers to health but also to improve health and health care. Efforts to determine when genetic tests are reliable enough for routine clinical use are quintessential public health activities.[32,33] The Secretary's Advisory Committee on Genetic Testing and its successor, the Secretary's Advisory Committee on Genetics, Health, and Society, were formed to provide such guidance.[34] The development of strategies to educate health care providers and patients about genomic medicine, a longstanding goal of the Human Genome Project, and to decrease obstacles to health-promoting behavior also falls comfortably within this rubric.

CONCLUSIONS

This brief discussion illustrates public expectations and fears about the effect of genomics, challenges to the goals of antidiscrimination laws and to the nature of the physician–patient relationship, and the contrasting perspectives and legal rules that apply to personal medical care and public health. Acknowledgment and examination of these complex issues are critical for identifying the appropriate ethical principles that should be applied and for creating the necessary legislative and regulatory responses.

Acknowledgments

Supported in part by a grant from the National Institutes of Health (5R01 HG 01974–02).

I am indebted to William O. Cooper and Jay Clayton for their helpful comments on earlier drafts and to Vanessa M. Spencer for her research assistance.

References

1. Welch CA. Sacred secrets—the privacy of medical records. N Engl J Med 2001; 345:371–2.
2. The Health Privacy Project. The state of health privacy: an uneven terrain (a comprehensive survey of state health privacy statutes). (Accessed July 15, 2003, at http://www.georgetown.edu/research/ihcrp/privacy/statereport.pdf.)
3. Gostin LO. National health information privacy: regulations under the Health Insurance Portability and Accountability Act. JAMA 2001; 285:3015–21.
4. Lapham EV, Kozma C, Weiss JO. Genetic discrimination: perspectives of consumers. Science 1996; 274:621–4.
5. Hall MA, Rich SS. Patients' fear of genetic discrimination by health insurers: the impact of legal protections. Genet Med 2000; 2:214–21.
6. Pokorski RJ. Insurance underwriting in the genetic era. Am J Hum Genet 1997; 60:205–16.
7. Calvo C, Johnson A, eds. Genetics policy report: insurance issues. Washington, D.C.: National Conference of State Legislatures, September 2001.
8. Genetic information and health insurance: enacted legislation as of April 29, 2002. (Accessed July 15, 2003, at http://www.nhgri.nih.gov/Policy_and_public_affairs/Legislation/insure.htm.)
9. Health Insurance Portability and Accountability Act of 1996, Pub. L. No. 104-191 (codified in part at 26 U.S.C. §§ 9801–9806, 29 U.S.C. §§ 1162–1167, 1181–1191c, and 42 U.S.C. § 300gg).
10. Health insurance law for workers changing jobs misses its goal. Wall Street Journal. March 12, 1998:B8.
11. Hudson KL, Rothenberg KH, Andrews LB, Kahn MJ, Collins FS. Genetic discrimination and health insurance: an urgent need for reform. Science 1995; 270:391–3.

12. Genetic Information Nondiscrimination in Health Act of 2003, S. 1053, 108th Cong., 1st Sess.

13. Chance PF. Overview of hereditary neuropathy with liability to pressure palsies. Ann N Y Acad Sci 1999; 883:14–21.

14. Lewin T. Commission sues railroad to end genetic testing in work injury cases. New York Times. February 10, 2001:A7.

15. Burlington Northern settles suit over genetic testing. New York Times. April 19, 2001:C4.

16. 42 U.S.C. §§ 12111–12117 (2001).

17. Calvo C, Johnson A, eds. Genetics policy report: employment issues. Washington, D.C.: National Conference of State Legislatures, September 2001.

18. Chevron, U.S.A., Inc. v. Echazabal, 122 S. Ct. 2045 (2002).

19. Toyota Motor Mfg., Inc. v. Williams, 122 S. Ct. 681 (2002).

20. Silvers A, Stein MA. An equality paradigm for preventing genetic discrimination. Vanderbilt Law Rev 2002; 55:1341–95.

21. Rothstein MA, Anderlik MR. What is genetic discrimination, and when and how can it be prevented? Genet Med 2001; 3:354–8.

22. Safer v. Pack, 677 A.2d 1188 (N.J. App.), appeal denied, 683 A.2d 1163 (N.J. 1996).

23. Clayton EW. What should the law say about disclosure of genetic information to relatives? J Health Care Law Policy 1998; 1:373–90.

24. Tarasoff v. Regents of University of California, 551 P.2d 334 (Cal. 1976).

25. Andrews LB, Fullarton JE, Holtzman NA, Motulsky AG, eds. Assessing genetic risks: implications for health and social policy. Washington, D.C.: National Academy Press, 1994.

26. Professional disclosure of familial genetic information. Am J Hum Genet 1998; 62:474–83.

27. President's Commission for the Study of Ethical Problems in Medicine and Biomedical and Behavioral Research. Screening and counseling for genetic conditions: a report on the ethical, social, and legal implications of genetic screening, counseling, and education programs. Washington, D.C.: Government Printing Office, 1983.

28. Pate v. Threlkel, 661 So. 2d 278 (Fla. 1995).

29. Fleisher LD, Cole LJ. Health Insurance Portability and Accountability is here: what price privacy? Genet Med 2001; 3:286–9.

30. Creason v. State Dept. Health Services, 957 P.2d 1323 (Cal. 1998).

31. Khoury MJ, McCabe LL, McCabe ERB. Population screening in the age of genomic medicine. N Engl J Med 2003; 348:50–8.

32. Holtzman NA, Watson MS, eds. Promoting safe and effective genetic testing in the United States: final report of the Task Force on Genetic Testing. Baltimore: Johns Hopkins University Press, 1998.

33. Centers for Disease Control and Prevention. Translating advances in human genetics into public health action: a strategic plan. (Accessed July 15, 2003, at http://www.cdc.gov/genomics/about/strategic.htm.)

34. Secretary's Advisory Committee on Genetic Testing. Enhancing the oversight of genetic tests: recommendations of the Secretary's Advisory Committee on Genetic Testing. (Accessed July 15, 2003, at http://www4.od.nih.gov/oba/sacgt.htm.)

14

Genomics as a Probe for Disease Biology

WYLIE BURKE, M.D., Ph.D.

Although our understanding of pathology has grown rapidly in recent decades, the underlying mechanisms of many diseases remain obscure. Genomic research offers a new opportunity for determining how diseases occur, by taking advantage of experiments of nature and a growing array of sophisticated research tools to identify the molecular abnormalities underlying disease processes.[1] In this chapter I examine examples in which genomic research has improved our understanding of molecular pathobiology and consider its potential for contributing to the study of common complex diseases.

EFFECT OF MUTATIONS ON THE SEVERITY OF HEMOPHILIA A

Before the advent of therapy for hemophilia A, some affected patients had only moderate bleeding problems, lived to adulthood, and led relatively normal lives in the absence of trauma or surgery.[2] Others had severe, spontaneous bleeding beginning in early childhood and rarely survived to adulthood; still others had disease of intermediate clinical severity. The main cause of this variation is the different mutations of the hemophilia A gene that cause hemophilia. When a mutation results in the complete loss of factor VIII protein—usually because a large gene deletion or genetic inversion results in the failure of gene transcription—severe hemophilia occurs.[2] The absence of endogenous factor VIII also increases the likelihood of an immune response when factor VIII–replacement therapy is used.[3] This immune response is a severe complication because it can prevent effective treatment. In contrast, other mutations, such as those in which there is a small change in the DNA sequence of the gene, lead to amino acid substitutions in the factor VIII protein. Depending on the nature of the substitution and its effect on the function of factor VIII, these mutations cause mild-to-moderate disease.

Originally published September 4, 2003

This example demonstrates that there is often a logical relation between a person's DNA sequence (genotype) and health outcome (phenotype). The functional effect of the genotype is the key factor, providing a molecular explanation for the severity of a given disease.

RECLASSIFICATION OF THE DYSTROPHINOPATHIES THROUGH GENOMIC UNDERSTANDING

Duchenne's muscular dystrophy is caused by mutations in the dystrophin gene—usually deletions or gene inversions—that produce total or near-total loss of the dystrophin protein from skeletal muscle, resulting in early and progressive loss of muscle function (Fig. 14.1).[5] Mutations in the dystrophin gene that cause less severe deficits in the final protein product result in Becker's muscular dystrophy, a milder disease that was historically considered a separate clinical entity. This disorder differs from Duchenne's muscular dystrophy in its later onset and milder course.[5]

When genomic research showed that these two clinically distinct disorders involved the same gene, a family of clinical disorders known as dystrophinopathies was identified.[4]

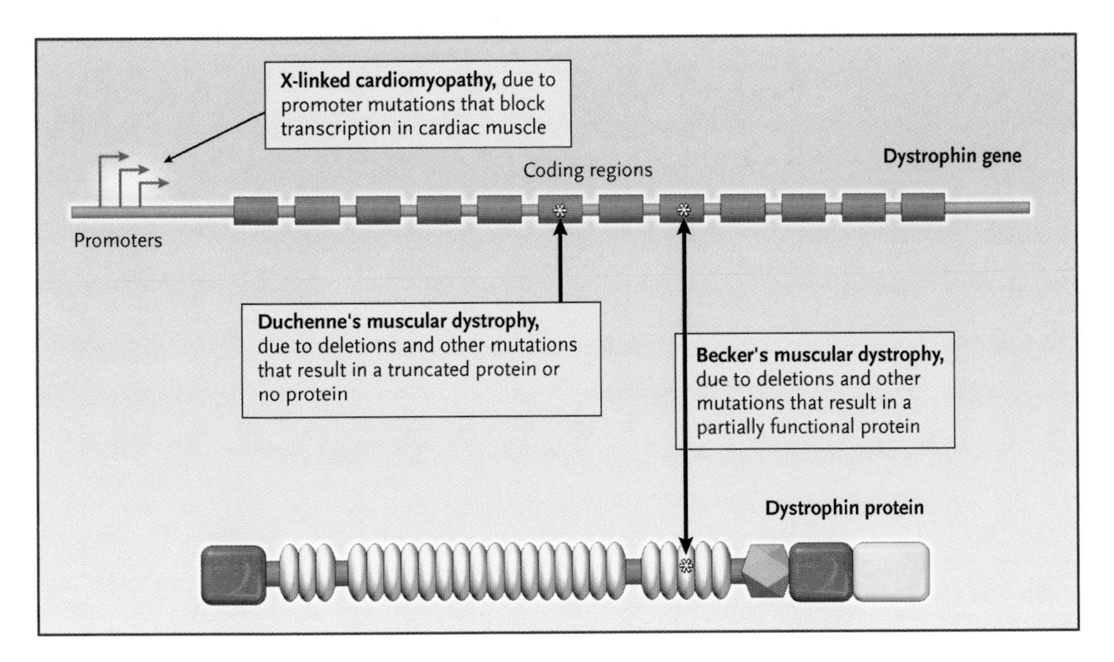

Figure 14.1 Clinical Effects of Various Mutations in the Dystrophin Gene. The dystrophin gene consists of promoter regions, regions coding for the dystrophin protein, and noncoding regions; in addition, smaller protein products of uncertain function are produced through the action of additional internal promoters and alternative splicing of transcripts.[1,4] The dystrophin protein consists of binding regions at either end and a central portion consisting of units known as spectrin-like repeats that are thought to give the protein a flexible, rod-like structure.[4] Mutations that result in either no protein product or a truncated protein product—usually large gene deletions or inversions—cause Duchenne's muscular dystrophy. Mutations that change the protein structure so as to partially alter its function result in the milder Becker's muscular dystrophy. X-linked dilated cardiomyopathy is caused by promoter mutations that result in the selective loss of dystrophin from heart-muscle tissue owing to the loss of transcription. Adapted from Blake et al.[4]

A third dystrophinopathy, X-linked dilated cardiomyopathy, was subsequently discovered. This disorder is caused by specific mutations in the dystrophin gene that lead to the selective loss of dystrophin from cardiac muscle, while dystrophin levels in skeletal muscles remain normal or nearly normal.[4,6] These mutations appear to affect one of the promoter regions of the gene, resulting in the selective loss of gene transcription in cardiac tissue (Fig. 14.1).[4,6] Discovery of the gene coding for dystrophin thus provided the means to understand a molecular relation among three seemingly different clinical disorders. As with hemophilia A, the relation between the genotype and the clinical outcome is the result of the functional effect of different mutations on dystrophin.

Another genetic disorder, hereditary hemorrhagic telangiectasia,[7] provides an interesting contrast. This disorder (also known as Osler–Weber–Rendu disease) causes vascular dysplasia, resulting in epistaxis, hemoptysis, and gastrointestinal bleeding. It is inherited as an autosomal dominant disorder and was assumed to be due to mutations in a single gene, but molecular studies revealed the involvement of two genes, one encoding activin-receptor–like kinase 1 *(ALK1)*[8] and one encoding endoglin.[9] Thus, the discovery of the genetic causes in this case revealed an unexpected complexity. Molecular studies also provided an explanation for the similar clinical outcome of mutations in two different genes: both protein products appear to function in the same or related biologic pathways.[8,9]

VARIABLE EFFECT OF THE SAME GENOTYPE ON HEALTH OUTCOMES IN CYSTIC FIBROSIS

Although the clinical effect of genetic mutations can often be predicted on the basis of their functional effect, the relation between the genotype and phenotype is sometimes less direct, as illustrated by cystic fibrosis. Cystic fibrosis results from mutations in the cystic fibrosis transmembrane conductance regulator *(CFTR)* gene; different mutations in the gene have various effects on the function of CFTR protein.[10,11] More severe mutations are associated with a loss of function, through defective synthesis, defective maturation, or blocked activation of the CFTR protein. Other mutations result in only partial loss of CFTR function, and as expected, these mutations are often associated with less severe disease,[10,11] including a later onset of symptoms, pancreatic sufficiency, and sometimes milder pulmonary disease.[3,12,13]

However, the correlation between genotype and phenotype in cystic fibrosis is imprecise. In particular, the severity of pulmonary disease cannot be predicted for most *CFTR* genotypes, including the most common one—ΔF508/ΔF508[11]—since both the age at onset of pulmonary symptoms and the rate of decline in pulmonary function vary.[14–16] For example, in some patients with the ΔF508/ΔF508 genotype, chronic respiratory symptoms do not develop until adolescence or adulthood and pulmonary function in young adulthood can range from highly compromised to normal.[11,15,16] This discrepancy is important, because pulmonary complications are typically the most serious and life-threatening manifestations of cystic fibrosis.

Both genetic and nongenetic factors appear to modify the effect of the cystic fibrosis genotype.[11] A Danish study found that certain genetic variants of mannose-binding lectin, a protein that functions in innate immune responses, were associated with greater loss of lung function in patients with cystic fibrosis.[17] Interestingly, this adverse effect was limited to patients with chronic *Pseudomonas aeruginosa* infection, indicating an interaction between the gene variant and the environment. Other studies, some of which used animal models of cystic fibrosis, provide additional evidence of the existence of genetic modifiers of the clinical effect of *CFTR* genotypes.[11,18–20] Some modifiers appear to be organ-specific; for example, a gene locus has been identified that influences whether an infant with cystic fibrosis will have neonatal meconium ileus, but it does not appear to influence the severity of lung disease.[21] Nongenetic modifiers of the cystic fibrosis phenotype, such as exposure to environmental tobacco smoke as a factor in the poor outcome of the disease[22] or exposure to respiratory pathogens as a contributor to the progression of lung disease, have been described.[10,11,23] The existence of environmental modifiers is also suggested by studies demonstrating a correlation between life expectancy and health insurance status[24] or socioeconomic status.[25] These findings could reflect the modifying effect of factors such as nutritional status or access to antibiotic treatment or could represent the effect of other environmental modifiers that have yet to be described.

Further evidence of the complexity of the genetics of cystic fibrosis comes from data on other phenotypes associated with *CFTR* mutations. Some mutations cause male infertility through the congenital absence of the vas deferens, without accompanying lung disease,[26] and others cause idiopathic chronic pancreatitis and mild, late-onset pulmonary disease.[11,27] Limited data suggest that *CFTR* mutations may also contribute to some cases of chronic sinusitis and allergic bronchopulmonary aspergillosis.[27] As we learn more about the function of CFTR and the effect of other genetic and nongenetic factors on this protein, we will most likely increase our understanding of the biology of both cystic fibrosis and related disorders.

GENETIC VARIANTS AS A COMMON PHENOMENON

Variants in many different genes form the basis for the genetic contribution to disease across the spectrum from rare disorders such as cystic fibrosis to common complex disorders such as cancer and heart disease. Genetic variants occur because new mutations arise at a low but continual rate in human tissues. Mutations that arise in germ-line tissues can subsequently be inherited, increasing the genetic variation in the population. The recently completed draft sequence of the human genome includes a catalog of 1.4 million single-nucleotide polymorphisms—sites where variations occur in the bases that form the building blocks of the DNA sequence.[28] Most DNA-sequence variations occur in noncoding regions of the genome—that is, in regions that do not code for protein products. Changes that occur in coding regions, however, can affect the function or efficiency of the protein that a gene encodes.

These differences can have physiological effects that are clinically important, such as causing differences in the response to drugs or environmental exposures or differences in susceptibility or the predisposition to diseases. For example, many of the enzymes responsible for drug metabolism occur in variant forms, leading to differences between people in the efficacy of a drug and the risk of adverse effects.[29,30] Similarly, many genetic variants that contribute to the risk of common diseases such as cancer and heart disease have been identified.[31] Unlike the gene changes that cause genetic disease, which tend to be rare and result in clinically significant loss of function, most common genetic variants cause relatively small changes in function.

Genetic changes may be adaptive in certain circumstances and harmful in others. An often-cited example is the association of the sickle cell trait—a mutation in the hemoglobin A gene—with resistance to malaria.[32] The selective advantage conferred by this resistance accounts for the high prevalence of the sickle trait in populations originating from regions where malaria is endemic. However, the presence of two copies of the sickle cell trait results in sickle cell anemia, a disease associated with multiple complications and premature mortality. In this case, genetics offered insights into disease biology that would not otherwise have been attainable. The Human Genome Project is likely to document many instances in which the effects of genetic variation are dependent on the context. As an example, a variant in the interleukin-1–receptor agonist, a protein that inhibits the inflammatory response, is associated with an increased risk of certain autoimmune diseases, but it may reduce the morbidity of some infections.[33]

ASTHMA AS AN EXAMPLE OF THE GENETICS OF COMMON COMPLEX DISEASES

Both genetic and environmental risk factors have an important role in most common diseases.[31] The same strategies developed for the study of genetic diseases can be applied to the study of common diseases, but the task is more difficult because complex patterns of gene–gene and gene–environment interactions must be evaluated. Asthma provides an illustration of this challenge.

Epidemiologic studies provide evidence that multiple genetic and environmental factors contribute to the causation of asthma, a clinical condition that is best viewed as a cluster of related disorders.[34] Patients with asthma vary with respect to the age at onset, course, sensitivity to specific environmental precipitants, and response to medications, and the relative contribution of genetic and nongenetic factors may also vary considerably among patients. Furthermore, the prevalence of asthma has risen dramatically in the past two decades, indicating that environmental risk factors have a key role.[35] Control of environmental risk factors and improved treatment are the primary public health strategies for the prevention of asthma.

Nevertheless, genetic factors contribute substantially to the risk of asthma.[34,36–38] How, in this context, might the study of the genetics of asthma contribute to better health care outcomes? There are several possibilities. A classification of asthma based on genetics might provide a more accurate means of defining clinical subtypes that benefited from specific treatments. Genetic classification might also provide improved prognostic information, including the identification of patients who are at highest risk for severe or life-threatening episodes of asthma. A better understanding of the molecular processes involved in the different pathways of asthma is also likely to lead to a more detailed understanding of the pathophysiology of the disease. This effort could lead to a more precise definition of the environmental modifications most likely to reduce the risk of asthma. It could also lead to improved drug treatment through genetic testing to predict a patient's responses to a drug or through the development of new drug therapies.[36,38]

Studies to identify genes associated with asthma use mapping techniques to pinpoint gene loci linked to asthma[1] and physiological studies to identify genes that are likely to affect the disease process. Both approaches have been productive. Mapping techniques have identified several genes associated with asthma.[37,38] One of these has yielded information about a metalloproteinase, ADAM-33, that may have a role in inflammatory responses or smooth-muscle hypertrophy and hyperreactivity.[39,40] Physiological studies have led to the characterization of genetic variants associated with asthma or atopic airway inflammation in several biologic pathways potentially related to asthma—for example, the beta-adrenergic receptor,[41] cytokines associated with the secretion of immunoglobulin E and airway inflammation,[42] and a transferase presumed to be involved in the detoxification of inhaled irritants.[43] One study reported a gene–environment interaction in which the effect of smoking on the risk of asthma was increased by a specific beta-adrenergic—receptor genotype.[44]

This research is still in the early stages and faces a number of technical problems that will also apply to the study of other common diseases. These include the need for standardized definitions of asthma phenotypes and intermediate biologic measures associated with the risk of asthma,[45–47] well-defined populations in unbiased studies with sufficient power to detect small effects,[48–50] and the concurrent measurement of both environmental and genetic risk factors.[51] Finally, any reported association between a genetic variant and disease risk cannot be considered established until the results of the study have been replicated.[52,53]

Determining associations between genes and diseases is just the first step in the translation of genomic research into clinical insights. This effort will require increasing attention to the study of the functions of proteins, or "proteomics,"[54] including the characterization of proteins identified as a result of genomic research. Of the estimated 30,000 to 35,000 genes in the human genome, approximately half code for unknown proteins.[28] With use of the genetic techniques that are now available, these proteins, their functions, and their interactions will be increasingly identifiable.[55] Analysis of the interaction between genetic and environmental effects in relevant biologic pathways, some (perhaps many) of which remain to be discovered, will form an important part of the study of common diseases such as asthma. As methods to achieve this goal are developed, they will provide a more complete and detailed picture of disease processes than has ever previously been possible.[56]

CONCLUSIONS

The study of gene mutations has provided a new model of pathophysiology in which the molecular causes of disease are illuminated by genetics. Evidence is now emerging of the complex interactions between genes and between genes and the environment in the causation of many diseases, and the study of these interactions represents the next important step in genomic research. Efforts to understand the molecular mechanisms that underlie common complex diseases will build on insights and strategies developed in the study of single-gene diseases. However, the scope of the analysis is far greater and will require continuing efforts to develop and improve molecular and informatic tools that allow the simultaneous analysis of many genetic variants and environmental risk factors. The success of genomic research to date[28] suggests that this ambitious research enterprise will also ultimately succeed.

References

1. Guttmacher AE, Collins FS. Genomic medicine—a primer. N Engl J Med 2002; 347:1512–20.

2. Hoyer LW. Hemophilia A. N Engl J Med 1994; 330:38–47.

3. Goodeve AC, Williams I, Bray GL, Peake IR. Relationship between factor VIII mutation type and inhibitor development in a cohort of previously untreated patients treated with recombinant factor VIII (Recombinate). Thromb Haemost 2000; 83:844–8.

4. Blake DJ, Weir A, Newey SE, Davies KE. Function and genetics of dystrophin and dystrophin-related proteins in muscle. Physiol Rev 2002; 82: 291–329.

5. Korf BR, Darras BT, Urion DK. Dystrophinopathies [includes: Duchenne muscular dystrophy (DMD), pseudohypertrophic muscular dystrophy, Becker muscular dystrophy (BMD), and X-linked dilated cardiomyopathy (XLDCM)]. Seattle: University of Washington, 2003. (Accessed August 8, 2003, at http://www.geneclinics.org/profiles/dbmd/details.html.)

6. Beggs AH. Dystrophinopathy, the expanding phenotype: dystrophin abnormalities in X-linked dilated cardiomyopathy. Circulation 1997; 95:2344–7.

7. Guttmacher AE, Marchuk DA, White RI Jr. Hereditary hemorrhagic telangiectasia. N Engl J Med 1995; 333:918–24.

8. Johnson DW, Berg JN, Baldwin MA, et al. Mutations in the activin receptor-like kinase 1 gene in hereditary haemorrhagic telangiectasia type 2. Nat Genet 1996; 13:189–95.

9. McAllister KA, Grogg KM, Johnson DW, et al. Endoglin, a TGF-beta binding protein of endothelial cells, is the gene for hereditary haemorrhagic telangiectasia type 1. Nat Genet 1994; 8:345–51.

10. Mickle JE, Cutting GR. Clinical implications of cystic fibrosis transmembrane conductance regulator mutations. Clin Chest Med 1998; 19:443–58.

11. Zielenski J. Genotype and phenotype in cystic fibrosis. Respiration 2000; 67:117–33.

12. Gan K-H, Veeze HJ, van den Ouweland AMW, et al. A cystic fibrosis mutation associated with mild lung disease. N Engl J Med 1995; 333:95–9.

13. De Braekeleer M, Allard C, Leblanc JP, Simard F, Aubin G. Genotype-phenotype correlation in cystic fibrosis patients compound heterozygous for the A455E mutation. Hum Genet 1997; 101:208–11.

14. The Cystic Fibrosis Genotype–Phenotype Consortium. Correlation between genotype and phenotype in patients with cystic fibrosis. N Engl J Med 1993; 329:1308–13.

15. Santis G, Osborne L, Knight RA, Hodson ME. Independent genetic determinants of pancreatic and pulmonary status in cystic fibrosis. Lancet 1990; 336: 1081–4.

16. Burke W, Aitken ML, Chen SH, Scott CR. Variable severity of pulmonary disease in adults with identical cystic fibrosis mutations. Chest 1992; 102:506–9.

17. Garred P, Pressler T, Madsen HO, et al. Association of mannose-binding lectin gene heterogeneity with severity of lung disease and survival in cystic fibrosis. J Clin Invest 1999; 104:431–7.

18. Rozmahel R, Wilschanski M, Matin A, et al. Modulation of disease severity in cystic fibrosis transmembrane conductance regulator deficient mice by a secondary genetic factor. Nat Genet 1996; 12:280–7. [Erratum, Nat Genet 1996; 13:129.]

19. Salvatore F, Scudiero O, Castaldo G. Genotype-phenotype correlation in cystic fibrosis: the role of modifier genes. Am J Med Genet 2002; 111:88–95.

20. Mekus F, Laabs U, Veeze H, Tummler B. Genes in the vicinity of CFTR modulate the cystic fibrosis phenotype in highly concordant or discordant F508del homozygous sib pairs. Hum Genet 2003; 112:1–11.

21. Zielenski J, Corey M, Rozmahel R, et al. Detection of a cystic fibrosis modifier locus for meconium ileus on human chromosome 19q13. Nat Genet 1999; 22:128–9.

22. Campbell PW III, Parker RA, Roberts BT, Krishnamani MR, Phillips JA III. Association of poor clinical status and heavy exposure to tobacco smoke in patients with cystic fibrosis who are homozygous for the F508 deletion. J Pediatr 1992; 120:261–4.

23. Dipple KM, McCabe ER. Modifier genes convert "simple" Mendelian disorders to complex traits. Mol Genet Metab 2000; 71:43–50.

24. Curtis JR, Burke W, Kassner AW, Aitken ML. Absence of health insurance is associated with decreased life expectancy in patients with cystic fibrosis. Am J Respir Crit Care Med 1997; 155:1921–4.

25. Britton JR. Effects of social class, sex, and region of residence on age at death from cystic fibrosis. BMJ 1989; 298:483–7.

26. Dork T, Dworniczak B, Aulehla-Scholz C, et al. Distinct spectrum of CFTR gene mutations in congenital absence of vas deferens. Hum Genet 1997; 100:365–77.

27. Noone PG, Knowles MR. "CFTR-opathies": disease phenotypes associated with cystic fibrosis transmembrane regulator gene mutations. Respir Res 2001; 2:328–32.

28. Lander ES, Linton LM, Birren B, et al. Initial sequencing and analysis of the human genome. Nature 2001; 409:860–921. [Errata, Nature 2001; 411:720, 412:565.]

29. Evans WE, McLeod HL. Pharmacogenomics—drug disposition, drug targets, and side effects. N Engl J Med 2003; 348:538–49.

30. Weinshilboum R. Inheritance and drug response. N Engl J Med 2003; 348:529–37.

31. Collins FS. Shattuck Lecture—medical and societal consequences of the Human Genome Project. N Engl J Med 1999; 341:28–37.

32. Ashley-Koch A, Yang Q, Olney RS. Sickle hemoglobin (HbS) allele and sickle cell disease: a HuGE review. Am J Epidemiol 2000; 151:839–45.

33. Witkin SS, Gerber S, Ledger WJ. Influence of interleukin-1 receptor antagonist gene polymorphism on disease. Clin Infect Dis 2002; 34:204–9.

34. Patino CM, Martinez FD. Interactions between genes and environment in the development of asthma. Allergy 2001; 56:279–86.

35. Mannino DM, Homa DM, Akinbami LJ, Moorman JE, Gwynn C, Redd SC. Surveillance for asthma—United States, 1980–1999. MMWR CDC Surveill Summ 2002; 51:1–13.

36. Palmer LJ, Silverman ES, Weiss ST, Drazen JM. Pharmacogenetics of asthma. Am J Respir Crit Care Med 2002; 165:861–6.

37. Cookson WO. Asthma genetics. Chest 2002; 121:Suppl:7S–13S.

38. Hakonarson H, Wjst M. Current concepts of the genetics of asthma. Curr Opin Pediatr 2001; 13:267–77.

39. Van Eerdewegh P, Little RD, Dupuis J, et al. Association of the ADAM33 gene with asthma and bronchial hyperresponsiveness. Nature 2002; 418:426–30.

40. Shapiro SD, Owen CA. ADAM-33 surfaces as an asthma gene. N Engl J Med 2002; 347:936–8.

41. Liggett SB. Polymorphisms of the beta2-adrenergic receptor and asthma. Am J Respir Crit Care Med 1997; 156:S156–S162.

42. Renauld JC. New insights into the role of cytokines in asthma. J Clin Pathol 2001; 54:577–89.

43. Spiteri MA, Bianco A, Strange RC, Fryer AA. Polymorphisms at the glutathione S-transferase, GSTP1 locus: a novel mechanism for susceptibility and development of atopic airway inflammation. Allergy 2000; 55:Suppl 61:15–20.

44. Wang Z, Chen C, Niu T, et al. Association of asthma with beta(2)-adrenergic receptor gene polymorphism and cigarette smoking. Am J Respir Crit Care Med 2001; 163:1404–9. [Erratum, Am J Respir Crit Care Med 2002; 166:775.]

45. Postma DS, Meijer GG, Koppelman GH. Definition of asthma: possible approaches in genetic studies. Clin Exp Allergy 1998; 28:Suppl 1:62–6.

46. Koppelman GH, Meijer GG, Postma DS. Defining asthma in genetic studies. Clin Exp Allergy 1999; 29:Suppl 4:1–4.

47. Ayres J. Severe asthma phenotypes: the case for more specificity. J R Soc Med 2001; 94:115–8.

48. Palmer LJ, Cookson WO. Using single nucleotide polymorphisms as a means to understanding the pathophysiology of asthma. Respir Res 2001; 2:102–12.

49. Weiss ST. Association studies in asthma genetics. Am J Respir Crit Care Med 2001; 165:2014–5.

50. Little J, Bradley L, Bray MS, et al. Reporting, appraising, and integrating data on genotype prevalence and gene-disease associations. Am J Epidemiol 2002; 156:300–10.

51. Khoury MJ, Yang Q. The future of genetic studies of complex human diseases: an epidemiologic perspective. Epidemiology 1998; 9:350–4.

52. Ioannidis JP, Ntzani EE, Trikalinos TA, Contopoulos-Ioannidis DG. Replication validity of genetic association studies. Nat Genet 2001; 29: 306–9.

53. Hirschhorn JN, Lohmueller K, Byrne E, Hirschhorn K. A comprehensive review of genetic association studies. Genet Med 2002; 4:45–61.

54. Childs B, Valle D. Genetics, biology and disease. Annu Rev Genomics Hum Genet 2000; 1:1–19.

55. Abbott A. A post-genomic challenge: learning to read patterns of protein synthesis. Nature 1999; 402:715–20.

56. Burley SK, Almo SC, Bonanno JB, et al. Structural genomics: beyond the Human Genome Project. Nat Genet 1999; 23:151–7.

15

Welcome to the Genomic Era

ALAN E. GUTTMACHER, M.D., and FRANCIS S. COLLINS, M.D., Ph.D.

To him who devotes his life to science, nothing can give more happiness than increasing the number of discoveries, but his cup of joy is full when the results of his studies immediately find practical applications. —Louis Pasteur

This book focuses on the ways in which the rapidly appearing tools of genomics have already begun to change the practice of medicine. In the previous chapter, for instance, Burke explores how genomics has started to improve our understanding of the biology of health and disease in ways that were never before possible.[1] Although the book demonstrates that genomics has indeed begun to change the practice of medicine, it catalogues only the birth of the genomic era and thus no more captures in detail the ultimate effect of genomic medicine than does the examination of a newborn foretell what the mature adult will be like.

If the genomic era can be said to have a precise birth date, it was on April 14, 2003, when the international effort known as the Human Genome Project put a close to the pregenomic era with its announcement (available at http://www.genome.gov/11006929) that it had achieved the last of the project's original goals, the complete sequencing of the human genome. The extent and pace of progress in genomics are suggested by the fact that this achievement occurred 11 days shy of the 50th anniversary of the publication of Watson and Crick's seminal description of the DNA double helix. If science, technology, and medicine have consistently demonstrated anything, it is that they proceed at an ever-quickening pace. That we have gone in the past 50 years from the first description of the structure of our DNA to its complete sequencing gives some indication of how much the impact of genomic medicine on the health care of today's neonates will increase by the time they turn 50 years of age.

However, it is not solely the next 50 years that will witness important advances in genomic medicine. Many such advances have already occurred. Indeed, one need look

Originally published September 4, 2003

no further than the pages of the *New England Journal of Medicine* to see potent additional examples of what has occurred recently: the use of genomics for the rapid identification of newly discovered pathogens such as that involved in the severe acute respiratory syndrome (SARS)[2,3]; the use of gene-expression profiling to assess prognosis and guide therapy, as in breast cancer[4]; the use of genotyping to stratify patients according to the risk of a disease, such as the long-QT syndrome[5] or myocardial infarction[6]; the use of genotyping to shed light on the response to certain drugs, such as antiepileptic agents[7]; and the use of genomic approaches in the design and implementation of new drug therapies, such as imatinib for the hypereosinophilic syndrome,[8] and to improve our understanding of the role of specific genes in the causation of common conditions, such as obesity.[9,10] During this same brief period, other notable genomics-based advances in our understanding of biology and of health have included the first comprehensive analysis of human chromosome 7,[11] clarification of the male-specific region of the human Y chromosome,[12] and the identification of the gene responsible for progeria.[13]

In recent months we have seen not only the promise of the genomic era with respect to medicine, but also its pitfalls. An example of the latter has been the revelation that confusion and misinformation have occasionally accompanied the counseling of persons who undergo screening for mutations in the cystic fibrosis transmembrane conductance regulator gene *(CFTR)*—the gene responsible for cystic fibrosis. Different mutations in *CFTR* have different effects, leading to a range of phenotypes. Proper interpretation of screening results demands an understanding of the clinical implications of specific genotypes. For example, the relatively common 5T variant leads to the phenotype of classic cystic fibrosis only when it is accompanied by the R117H mutation on the same chromosome arm. However, there have been anecdotal reports of persons who were told that the presence of the 5T variant alone was indicative of a serious risk of cystic fibrosis. There is reason to believe that such problems are not intrinsic to genomic medicine and merely reflect the temporary difficulties of integrating virtually any new form of technology into health care. But this example certainly points out the urgency with which genetic literacy must be achieved among all health care providers. In the genomic age, *primum non nocere* remains a useful aim.

Genomics provides powerful means of discovering hereditary factors in disease. But even in the genomic era, it is not genes alone but the interplay of genetic and environmental factors that determines phenotype (i.e., health or disease). This point is not new, but it bears repeating. For example, a mutation in *CHE1* may be innocuous until a person carrying it is exposed to succinylcholine chloride anesthesia, when it leads to prolonged apnea. Conversely, substantial mutations in phenylalanine hydroxylase inevitably result in the sequelae of classic phenylketonuria, including profound mental retardation, unless the affected newborn is placed on a special diet, in which case essentially normal intellectual development can be expected. Recent reports further expand our knowledge of the complex interactions between genes and the environment. For instance, one such study suggests that common variations in the serotonin-transporter gene influence the likelihood of depression after exposure to stress.[14]

Since it remains difficult to alter genes in humans (for both technical and ethical reasons), for the next couple of decades we will generally use personalized modifications of the environment, and not of genes, to translate genomics-based knowledge into improvements in health for most of our patients. Clinicians will much more frequently suggest to patients with hereditary hemochromatosis that they avoid iron supplementation than that they consider gene therapy. Women who carry mutations in *BRCA2* will profit more from taking tamoxifen than from manipulations of their genotype.

With the end of the pregenomic era in sight, more than 600 experts recently collaborated to produce a vision for the future of genomic research and its applications to biology, health, and society.[15] According to that vision, for instance, within a decade or two, it will be possible to sequence anyone's entire genome for a laboratory cost of less than $1,000. If this proves true, one can imagine how not only research, but also clinical care, may change dramatically. However, as is true for so much of the application of genomics, ethical, legal, and social issues complicate this optimistic picture. Unless complex issues regarding the patenting and licensing of gene-based knowledge and techniques are dealt with more successfully than they are today, the "$1,000 genome" will remain a wish, not a reality. Even if the intellectual-property discussions are complex, the math is simple. Assuming that roughly half of the approximately 30,000 human genes are patented, if each patent holder were to charge only $1 per test to license his or her gene, the $1,000 genome would become the $16,000 genome, a very different economic, and thus clinical, reality.

Another social issue, with particular relevance in the United States, is the understandable concern of many patients that obtaining genetic information important to their health care is not worth the risk of discrimination stemming from the use of such information by potential insurers or employers. Although more than 40 states limit employers' and insurers' access to or use of genetic information, many people believe that only the passage of legislation mandating uniform national protection against the misuse of such information will lead to full use of genetic testing. Congressional leaders from both major parties and the current administration have supported such federal legislation, and passage of such legislation is currently closer to reality than ever before. However, until it is enacted and signed into law, the fear of discrimination will remain.

Other social issues require our attention if genomic medicine is to benefit our patients. How should genetic tests be regulated? What, if any, are the appropriate uses of direct-to-consumer marketing of genetic tests? The Internet has recently had a proliferation of genetic-testing sites that feature claims grounded in greed and pseudoscience, rather than in data or reality. How will health care providers and the public distinguish between these and responsible testing services, whether they are available through the Internet or in the hospital?

It would be easy to assume that for the forseeable future the benefits of genomic medicine will accrue only to people in developed countries. However, even in resource-poor regions of the world, genomic approaches can offer dramatic benefits to health, as the publication of the genomes for *Plasmodium falciparum*[16] and *Anopheles gambiae*[17] exemplifies. Nonetheless, another important social issue is the challenge of harnessing this unprecedented opportunity so that genomic medicine benefits all.

While recognizing such challenges, we look forward with curiosity and real hope to the advances of the next 50 years—the first 50 years of the genomic era. As evidenced by this book, today's researchers and clinicians have already started to use the power of genomics to improve health, and we anticipate that this is but a hint of the progress to come.

References

1. Burke W. Genomics as a probe for disease biology. N Engl J Med 2003; 349:969–74.

2. Ksiazek TG, Erdman D, Goldsmith CS, et al. A novel coronavirus associated with severe acute respiratory syndrome. N Engl J Med 2003; 348:1953–66.

3. Drosten C, Günther S, Preiser W, et al. Identification of a novel coronavirus in patients with severe acute respiratory syndrome. N Engl J Med 2003; 348:1967–76.

4. van de Vijver MJ, He YD, van 't Veer LJ, et al. A gene-expression signature as a predictor of survival in breast cancer. N Engl J Med 2002; 347:1999–2009.

5. Priori SG, Schwartz PJ, Napolitano C, et al. Risk stratification in the long-QT syndrome. N Engl J Med 2003; 348:1866–74.

6. Yamada Y, Izawa H, Ichihara S, et al. Prediction of the risk of myocardial infarction from polymorphisms in candidate genes. N Engl J Med 2002; 347: 1916–23.

7. Siddiqui A, Kerb R, Weale ME, et al. Association of multidrug resistance in epilepsy with a polymorphism in the drug-transporter gene *ABCB1*. N Engl J Med 2003; 348:1442–8.

8. Cools J, DeAngelo DJ, Gotlib J, et al. A tyrosine kinase created by fusion of the *PDGFRA* and *FIP1L1* genes as a therapeutic target of imatinib in idiopathic hypereosinophilic syndrome. N Engl J Med 2003; 348:1201–14.

9. Farooqi IS, Keogh JM, Yeo GSH, Lank EJ, Cheetham E, O'Rahilly S. Clinical spectrum of obesity and mutations in the melanocortin 4 receptor gene. N Engl J Med 2003; 348:1085–95.

10. Branson R, Potoczna N, Kral JG, Lentes K-U, Hoehe MR, Horber FF. Binge eating as a major phenotype of melanocortin 4 receptor gene mutations. N Engl J Med 2003; 348:1096–103.

11. Hillier LW, Fulton RS, Fulton LA, et al. The DNA sequence of human chromosome 7. Nature 2003; 424:157–64.

12. Skaletsky H, Kuroda-Kawaguchi T, Minx PJ, et al. The male-specific region of the human Y chromosome is a mosaic of discrete sequence classes. Nature 2003; 423:825–37.

13. Erikkson M, Brown WT, Gordon LB, et al. Recurrent de novo point mutations in lamin A cause Hutchinson-Gilford progeria syndrome. Nature 2003; 423:293–8.

14. Caspi A, Sugden K, Moffitt TE, et al. Influence of life stress on depression: moderation by a polymorphism in the 5-HTT gene. Science 2003; 301:386–9.

15. Collins FS, Green ED, Guttmacher AE, Guyer MS. A vision for the future of genomics research. Nature 2003; 422:835–47.

16. Gardner MJ, Hall N, Fung E, et al. Genome sequence of the human malaria parasite Plasmodium falciparum. Nature 2002; 419:498–511.

17. Holt RA, Subramanian GM, Halpern A, et al. The genome sequence of the malaria mosquito Anopheles gambiae. Science 2002; 298:129–49.

Index

Page numbers in *italics* denote figures; those followed by t denote tables.